TI-83 Technology Resource Manual

Bill Ardis
Collin County Community College

To Accompany

Mathematics
An Applied Approach
Eighth Edition

Michael Sullivan
Chicago State University

Abe Mizrahi
Indiana University Northwest

JOHN WILEY & SONS, INC.

Cover Photo: ©Photodisc/Getty Images

To order books or for customer service call 1-800-CALL-WILEY (225-5945).

ISBN 0-471-44824-9

Printed in the United States of America

10 9 8 7 6 5 4 3 2 1

Printed and bound by Odyssey Press, Inc.

Table of Contents

Preface

About This Manual

This manual was created with Microsoft Word 2002 using the TI-83 key font and MathType 5. All screen shots were captured using TI Connect.

Thank You

I would like to thank Michael Boezi for the opportunity to work on this manual, Jennifer Battista for her help and oversight on this manual; Kathleen Miranda and Kelly Boyle for their help; and the rest of the people at John Wiley & Sons for their support. Thanks to Michael Divinia for proof-reading the manuscript, your suggestions were greatly appreciated. Thanks again to Laurie Rosatone for getting me started doing the writing thing. Finally, thanks to my wife Ladan for her patience, understanding, and support.

Chapter 0 – How To Use This Manual

Introduction

This graphing calculator manual accompanies *Mathematics: An Applied Approach* 8 ed. by Sullivan & Mizrahi. The manual contains the solutions to the graphing calculator problems included in the exercises throughout the text. These problems are denoted by the calculator icon.

Please note that not every section in the textbook included graphing calculator problems in the exercise sets. If a particular section in missing in the manual it is because there were no calculator problems in the exercise set for that section.

Keep this manual handy as you work through the exercises. This manual provides a resource for the commands necessary to solve the graphing calculator problems. If you want more information on a particular command, refer to your calculator manual.

Notation for Keys Used in Manual

As you read through the solutions, please note that keystrokes are notated like ENTER or like [ANGLE]. A keystroke like ENTER refers to an operation on one of the keys, while a keystroke like [ANGLE] refers to an operation or menu that is above one of the keys. If the operation is in yellow, it must be preceded by the 2nd key. If the operation or character is in green, it must be preceded by the ALPHA key.

When a new key or function is first introduced, a brief explanation of the key or function, along with the general format are included, as well as the specific keystrokes necessary for solving the particular problem. On those problems where keystrokes are provided, match your results with the screens shown in the manual at each step.

Afterwards, only screen shots are included. The screen shots will include commands used, as well as the results, for each problem.

Chapter 1 – Linear Equations

Section 1.1 Rectangular Coordinates; Lines

In Problems 85-92, use a graphing utility to graph each linear equation. Be sure to use a viewing rectangle that shows the intercepts. Then locate each intercept rounded to two decimal places.

85. $1.2x + 0.8y = 2$

Graphing an equation on the TI-83 Plus is a three step process. You must tell the calculator what equation to graph, what part of the graph you want to see, and then have the calculator draw the graph. In order to enter the equation, you must first solve the equation for the dependent variable (in most cases y). When entering the equation into the calculator, use x as the independent variable. The viewing rectangle (or window) is the portion of the graph that the calculator will display (See Figure 1).

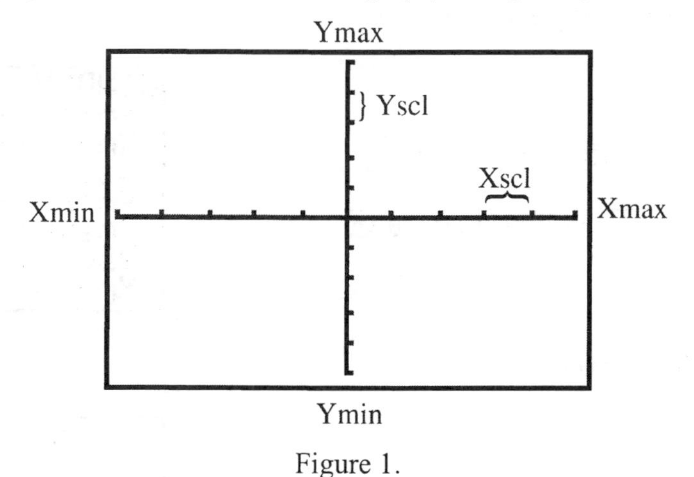

Figure 1.

One commonly used viewing window is the standard window, given in Figure 2 on the next page.

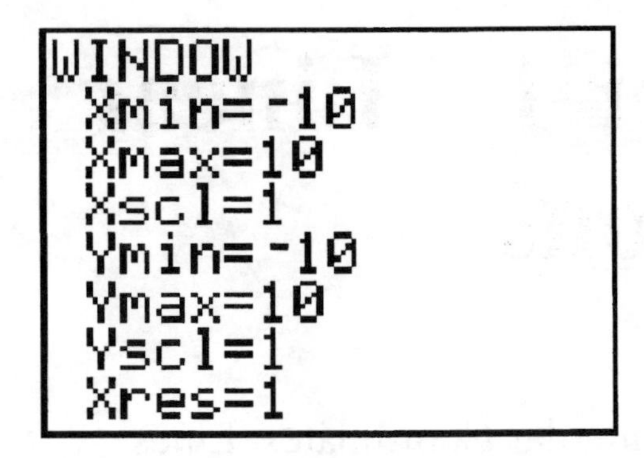

Figure 2.

First we must solve the equation for y. In this problem, we obtain $y = -1.5x + 2.5$.

[Y=] [(-)] [1] [.] [5] [X,T,Θ,n] [+] [2] [.] [5] [ENTER]

Plot1 Plot2 Plot3
\Y1=-1.5X+2.5
\Y2=
\Y3=
\Y4=
\Y5=
\Y6=
\Y7=

Next, set a viewing rectangle (or window).

[WINDOW] [(-)] [1] [0] [ENTER] [1] [0] [ENTER] [1] [ENTER]

[(-)] [1] [0] [ENTER] [1] [0] [ENTER] [1] [ENTER]

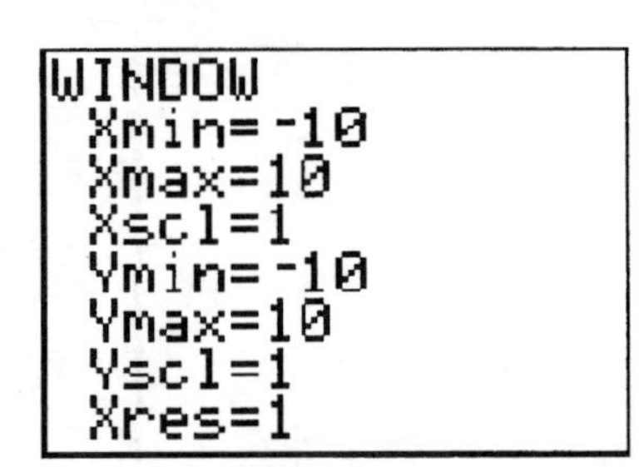

Finally, graph the equation.

[GRAPH]

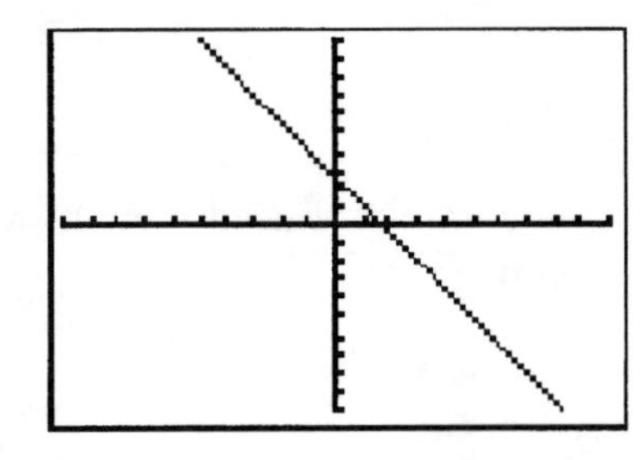

You can use your TI-83 Plus to find the x- and y-intercepts. The `value` function will find the value of a function at a given point. To find the y-intercept, we must find the `value` of the function when $x = 0$. The `zero` function will find the zeros (an x-intercept is a zero) of a function. The `zero` finder requires three inputs, an x-value to the left of the zero (x-intercept), an x-value to the right of the zero (x-intercept), and an estimate of the x-value of the zero (x-intercept). Both `value` and `zero` are found under the `[CALC]` menu.

Find the y-intercept.

[2nd] [TRACE] [1] [0] [ENTER]

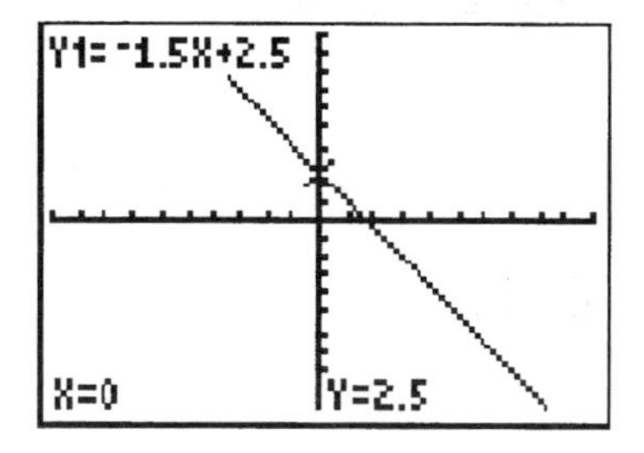

Find the x-intercept. Enter a value for x that is less than (to the left of) the x–intercept. Notice that the x-intercept is between $x = 1$ and $x = 2$, so we can use $x = 1$ as a left bound.

[2nd] [TRACE] [2] [1]

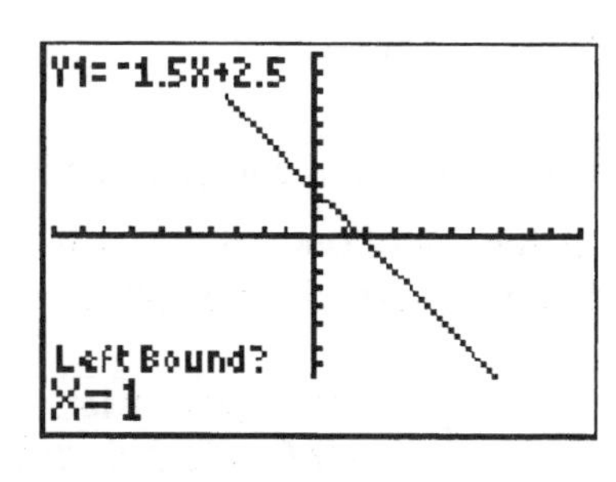

Input $x = 2$ as a right bound.

[ENTER] [2]

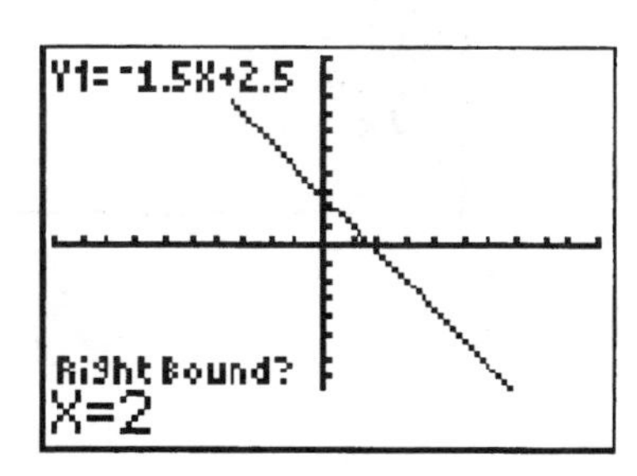

Input $x = 1.5$ as a guess.

[ENTER] [1] [.] [5]

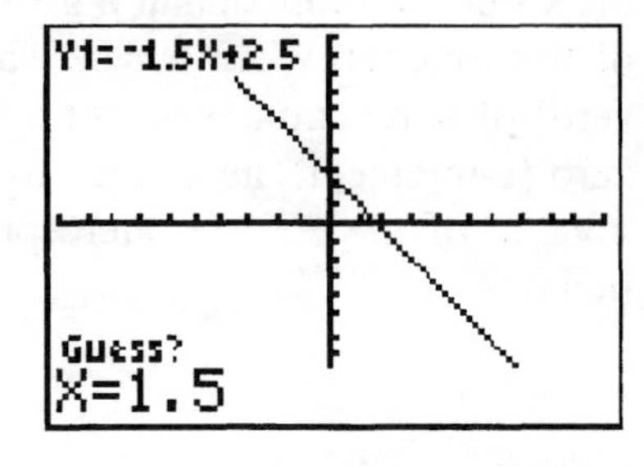

Find the x-intercept.

[ENTER]

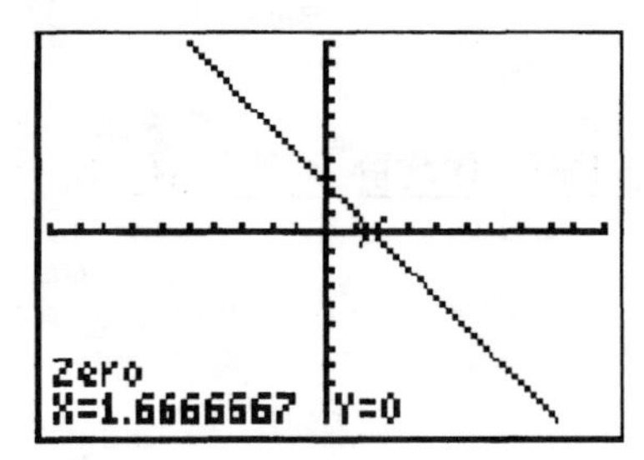

Thus the y-intercept is $(0, 2.50)$ and the x-intercept is approximately $(1.67, 0)$.

87. $21x - 15y = 53$

First, solve the equation for y. In this problem we obtain $y = \frac{7}{5}x - \frac{53}{15}$. Enter the equation in the function editor. Then, use a standard window and graph. The results are shown below and on the next page.

```
Plot1 Plot2 Plot3
\Y1=(7/5)X-53/15
\Y2=
\Y3=
\Y4=
\Y5=
\Y6=
```

```
WINDOW
 Xmin=-10
 Xmax=10
 Xscl=1
 Ymin=-10
 Ymax=10
 Yscl=1
 Xres=1
```

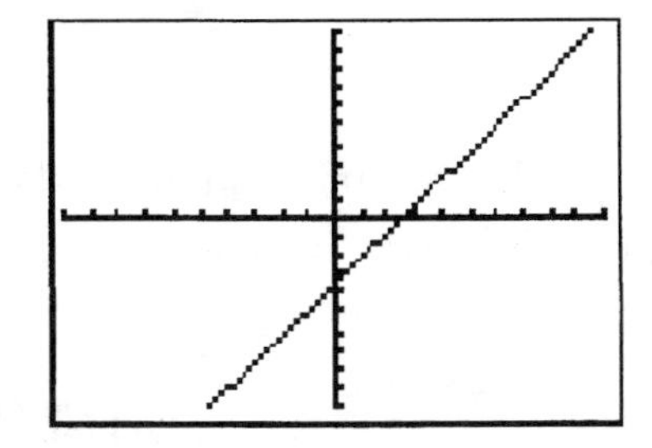

Use `zero` and `value` to find the x- and y-intercepts, respectively. Note that the x-intercept is between $x = 2$ and $x = 3$.

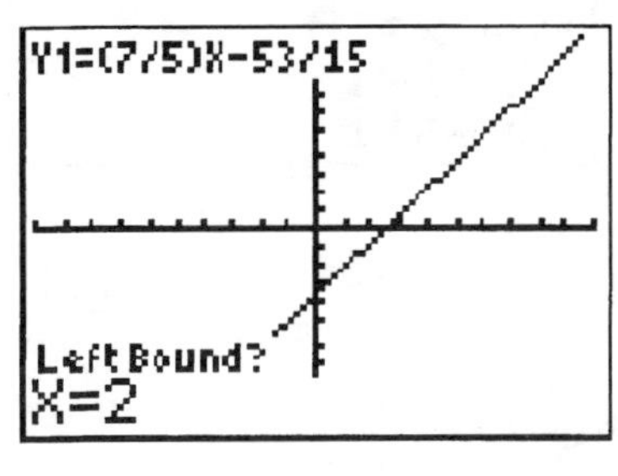

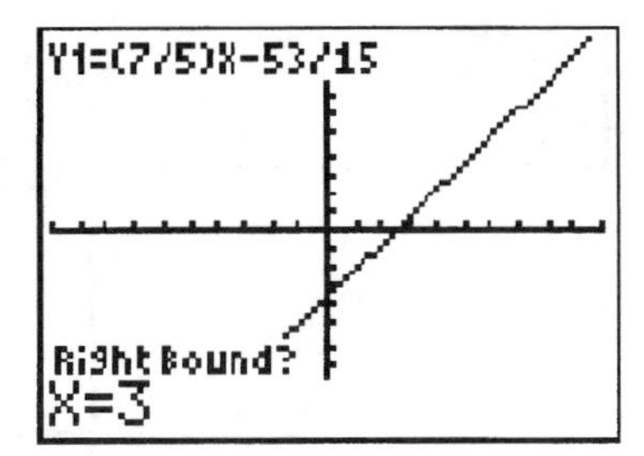

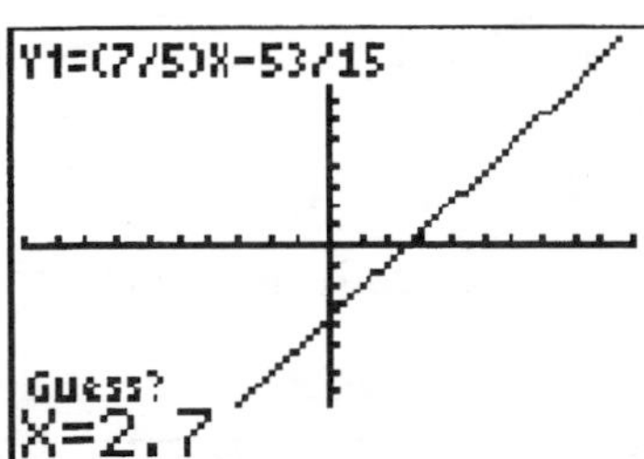

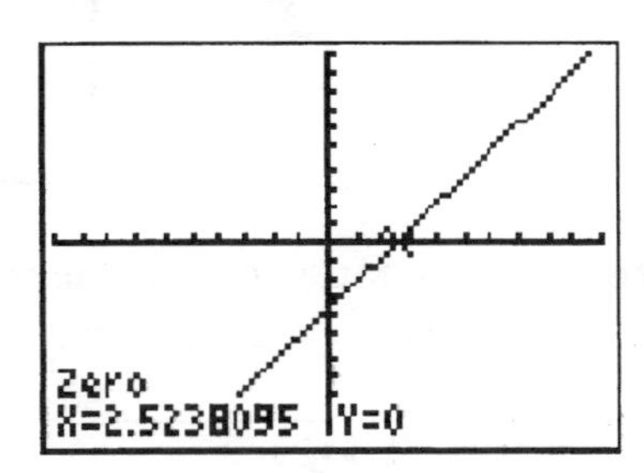

The x-intercept is approximately $(2.52, 0)$.

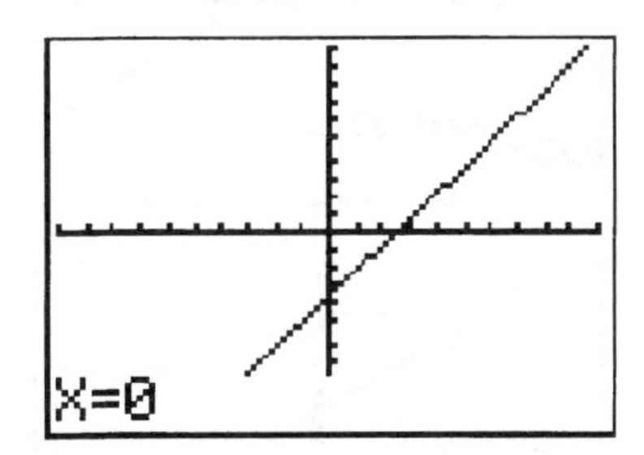

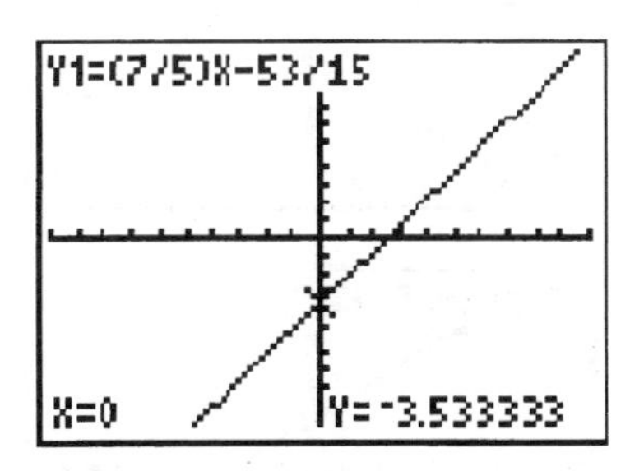

The y-intercept is approximately $(0, -3.53)$.

89. $\frac{4}{17}x+\frac{6}{23}y=\frac{2}{3}$

First, solve the equation for *y*. In this problem we obtain $y=-\frac{46}{51}x+\frac{23}{9}$. Enter the equation in the function editor. Then use a standard window and graph. The results are shown below.

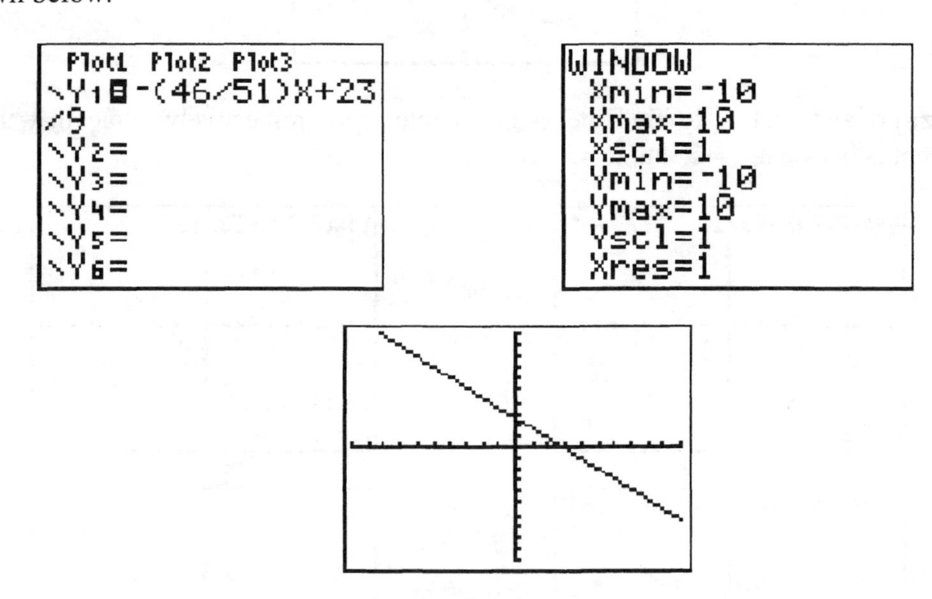

Use `zero` and `value` to find the *x*- and *y*-intercepts, respectively. Note that the *x*-intercept is between $x=2$ and $x=3$.

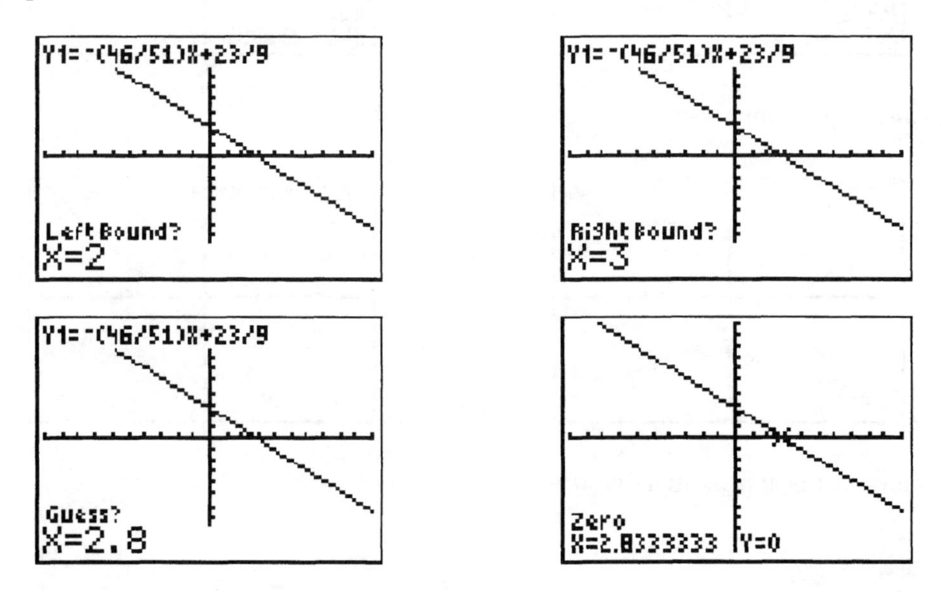

The *x*-intercept is approximately $(2.83,0)$.

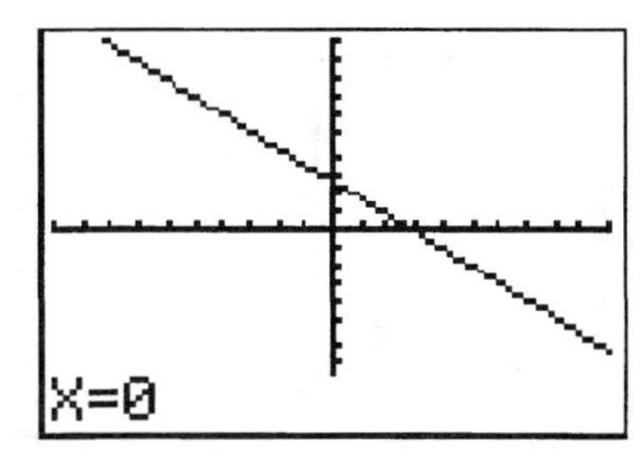

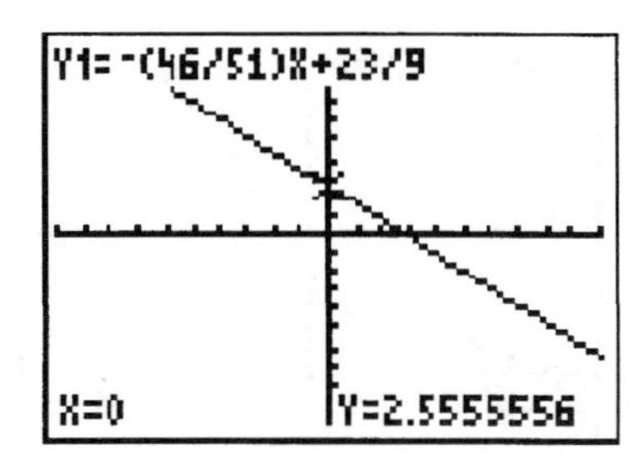

The y-intercept is approximately $(0, 2.56)$.

91. $\pi x - \sqrt{3}y = \sqrt{6}$

First, solve the equation for y. In this problem we obtain $y = \frac{\pi}{\sqrt{3}}x - \sqrt{2}$. Enter the equation in the function editor. Then use a standard window and graph. The results are shown below.

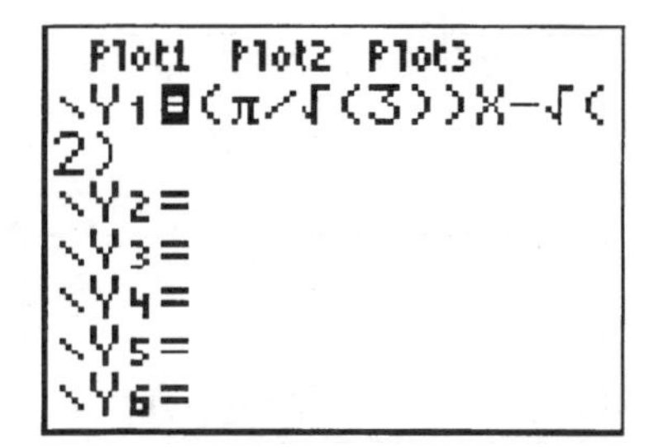

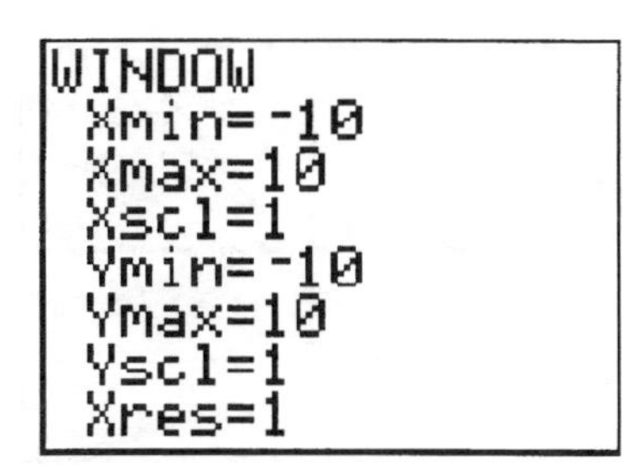

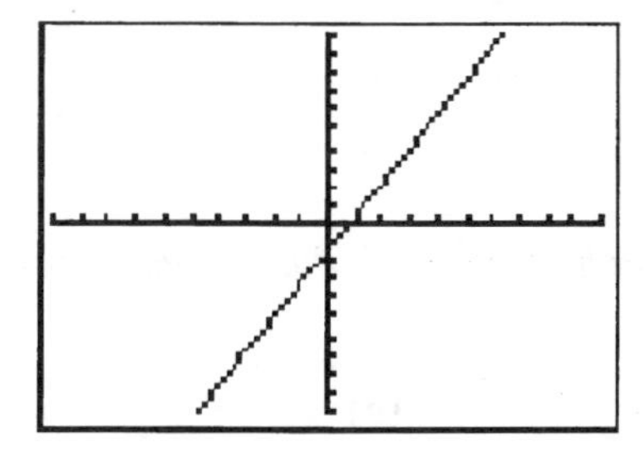

Use `zero` and `value` to find the x- and y-intercepts, respectively. Note that the x-intercept is between $x = 0$ and $x = 2$.

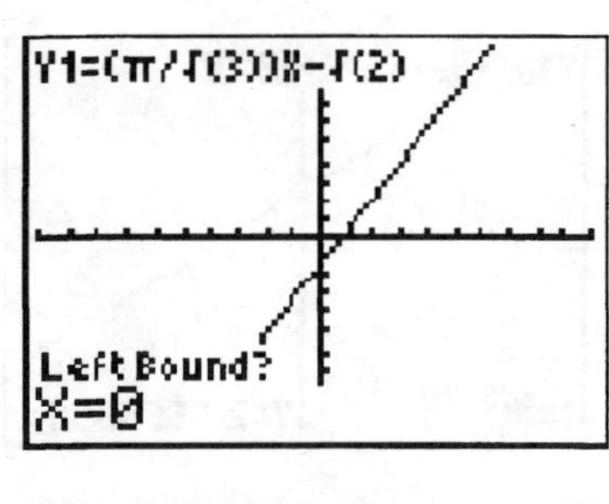

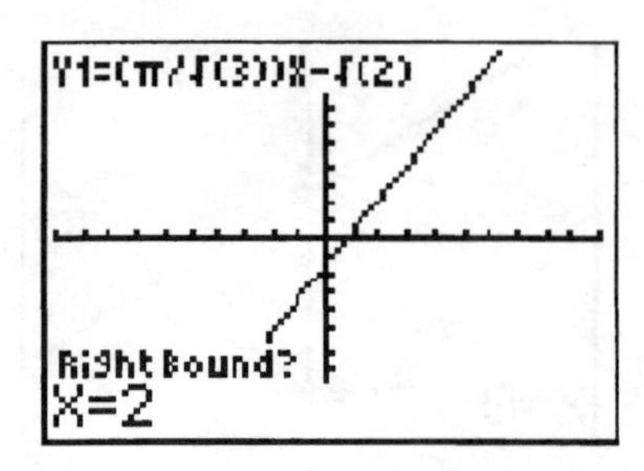

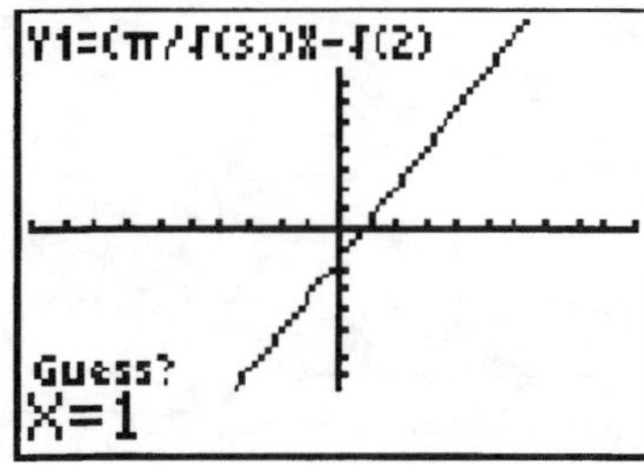

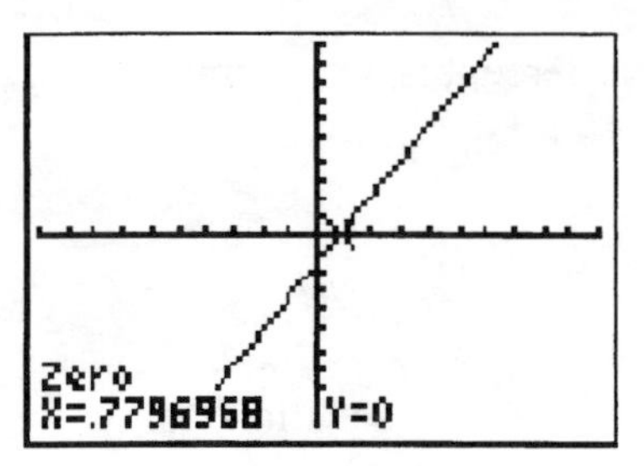

The x-intercept is approximately $(0.78, 0)$.

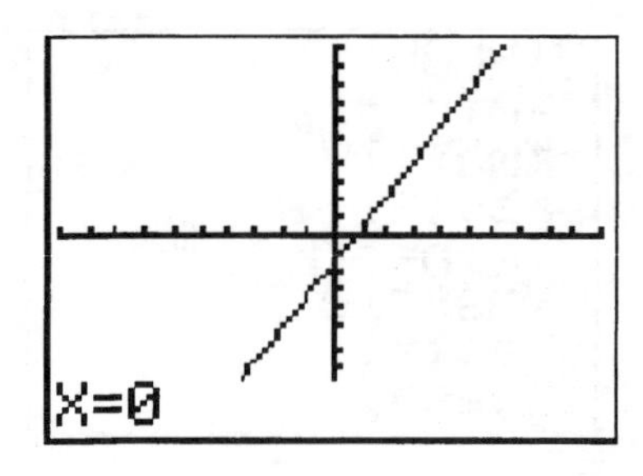

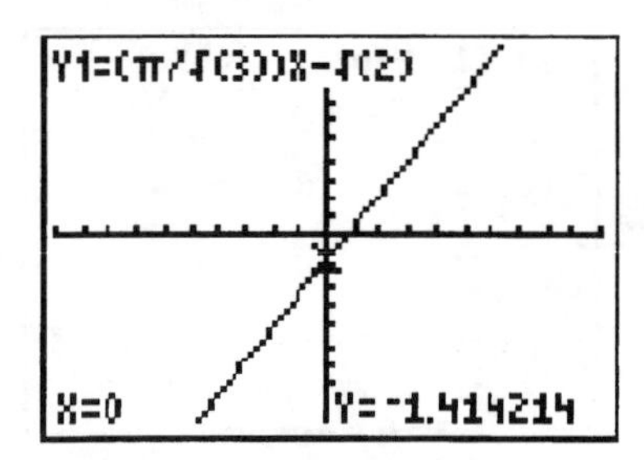

The y-intercept is approximately $(0, -1.41)$.

In Problems 93-96, match each graph with the correct equation:

(a) $y = x$; (b) $y = 2x$; (c) $y = \frac{x}{2}$; (d) $y = 4x$

93.

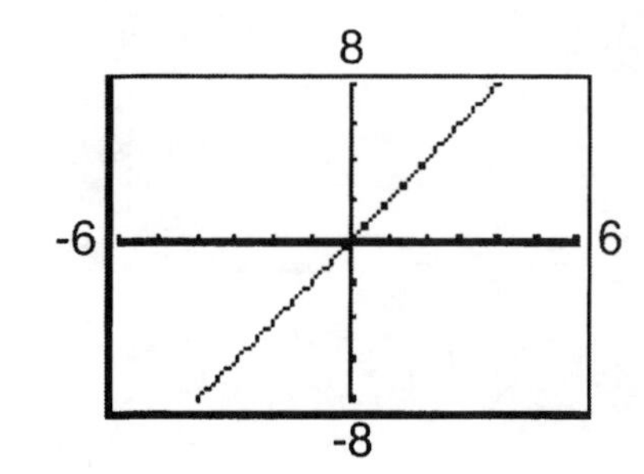

Enter all four equations in the function editor and use the window given in the problem.

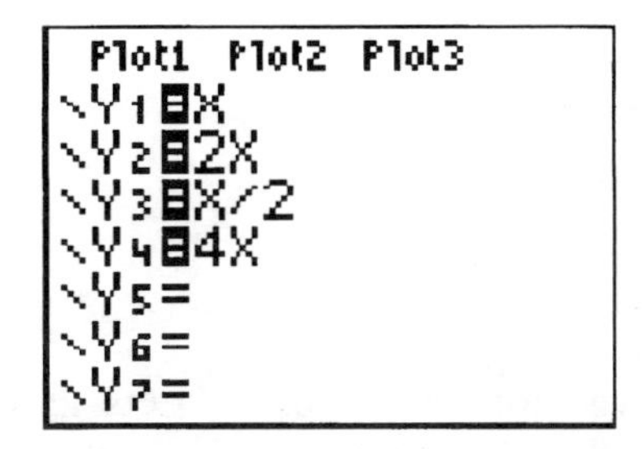

WINDOW
Xmin=-6
Xmax=6
Xscl=1
Ymin=-8
Ymax=8
Yscl=2
Xres=1

You can tell you calculator not graph an equation by "turning it off." To turn off an equation, return to the function editor, move the cursor to the equals sign and press [ENTER]. We will graph one equation at a time until we find a graph that matches the given graph.

Turn off the second equation.

[Y=] [▼] [◄] [ENTER]

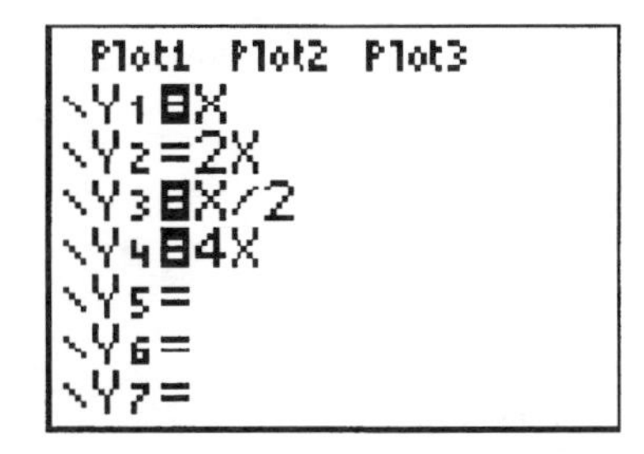

Turn off the third and fourth equations.

[▼] [ENTER] [▼] [ENTER]

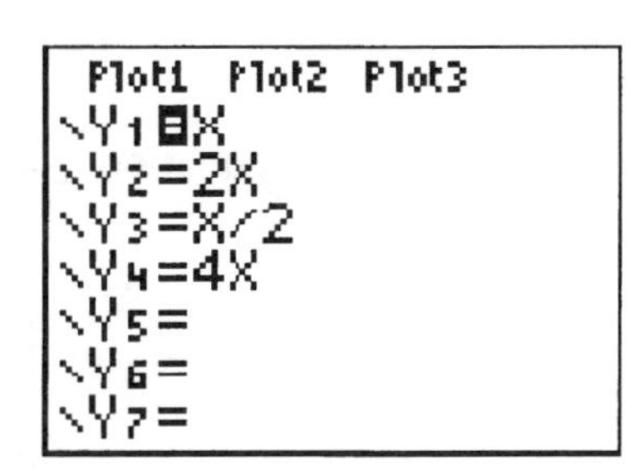

Graph the first equation $y = x$.

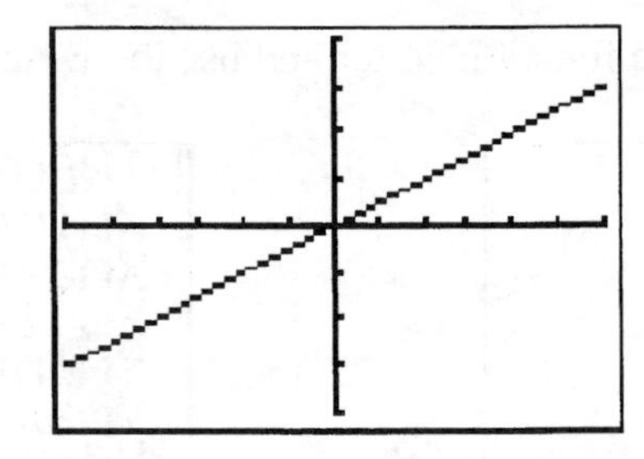

This graph does not match the graph we were given for this problem. Lets graph the second equation $y = 2x$. We must turn the first equation "off" and turn the second equation "on" to graph $y = 2x$.

[Y=] [◄] [ENTER]

```
Plot1 Plot2 Plot3
\Y1=X
\Y2=2X
\Y3=X/2
\Y4=4X
\Y5=
\Y6=
\Y7=
```

[▼] [ENTER]

```
Plot1 Plot2 Plot3
\Y1=X
\Y2■2X
\Y3=X/2
\Y4=4X
\Y5=
\Y6=
\Y7=
```

Graph the second equation $y = 2x$.

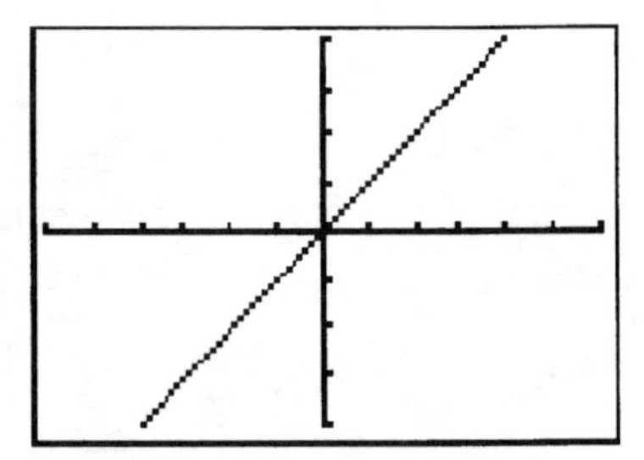

The graph is the graph of the equation (b) $y = 2x$.

95. 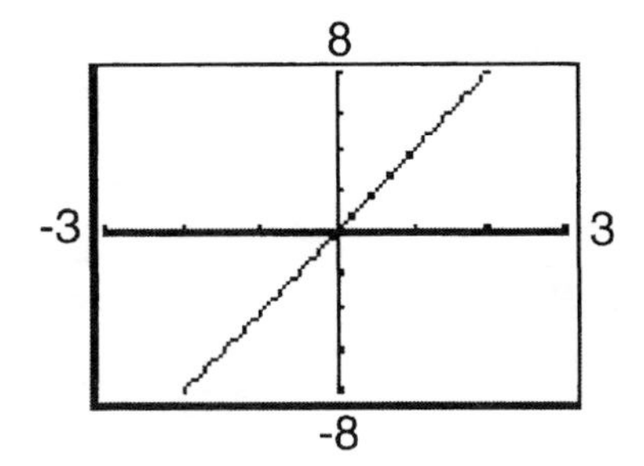

Enter all four equations in the function editor and use the window given in the problem. Use the steps discussed in Problem 93 to graph each equation individually.

```
WINDOW
 Xmin=-3
 Xmax=3
 Xscl=1
 Ymin=-8
 Ymax=8
 Yscl=4
 Xres=1
```

By process of elimination, we obtain the following graph.

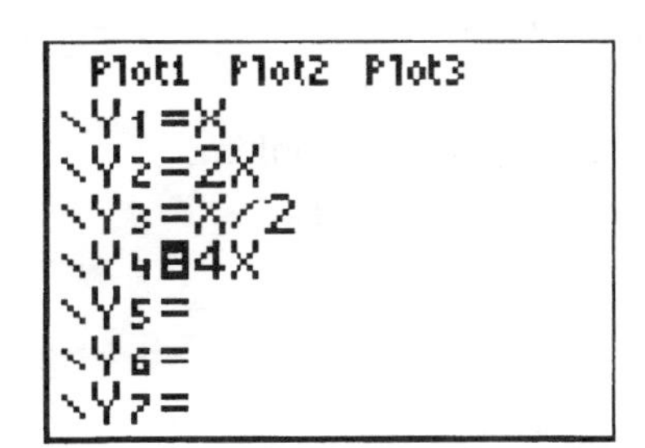

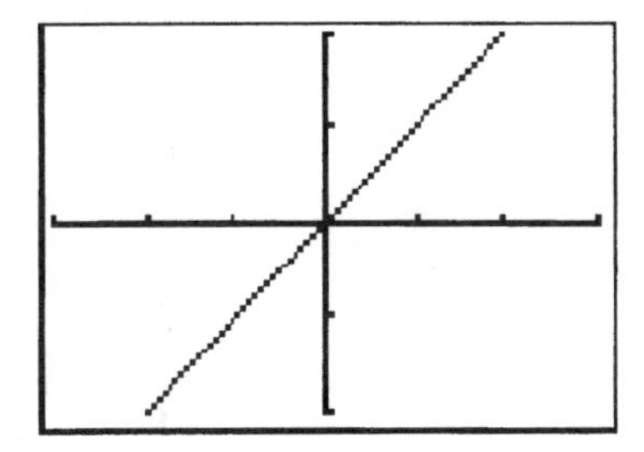

The graph is the graph of the equation (d) $y = 4x$.

In Problems 97–100, write an equation of each line. Express your answers using either the general form or the slope-intercept form of the equation of a line, whichever you prefer.

97.

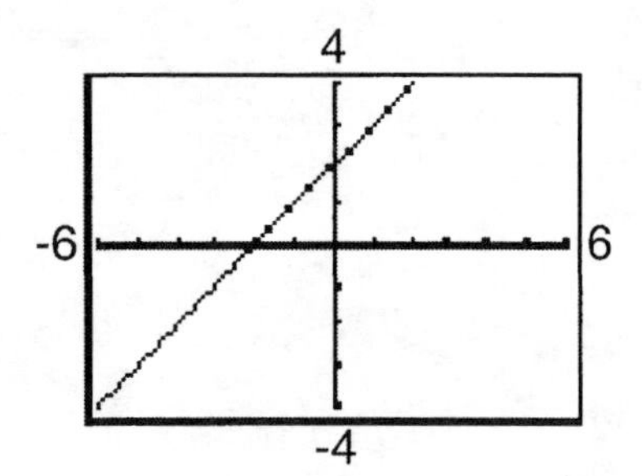

The x-intercept of the graph is $(-2,0)$ and the y-intercept is $(0,2)$. The slope of the line through these two points is

$$m=\frac{2-0}{0-(-2)}=\frac{2}{2}=1$$

Using the point slope form of the line we obtain

$$y-2=1(x-0) \text{ or } y-0=1(x-(-2))$$

Simplifying either equation we obtain

$$y=x+2 \text{ or } x-y=-2$$

We can check by graphing our result on the window given in the problem.

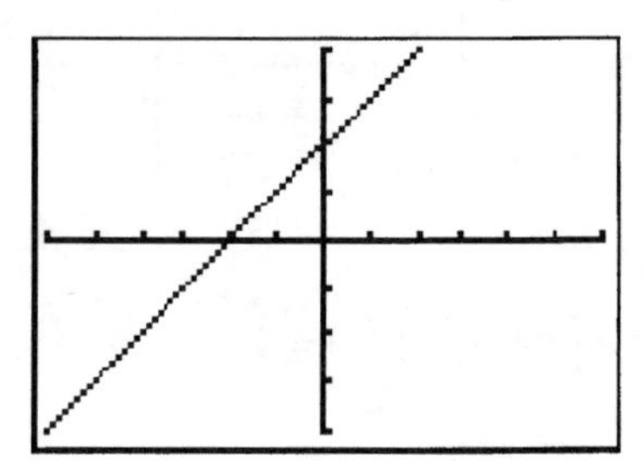

99.

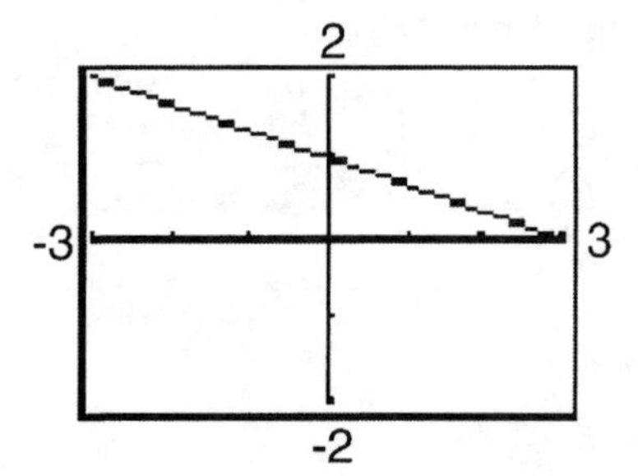

The x-intercept of the graph is $(3,0)$ and the y-intercept is $(0,1)$. The slope of the line through these two points is

$$m = \frac{1-0}{0-3} = \frac{1}{-3} = -\frac{1}{3}$$

Using the point slope form of the line we obtain

$$y-1 = -\tfrac{1}{3}(x-0) \text{ or } y-0 = -\tfrac{1}{3}(x-3)$$

Simplifying either equation we obtain

$$y = -\tfrac{1}{3}x+1 \text{ or } x+3y=1$$

We can check by graphing our result on the window given in the problem.

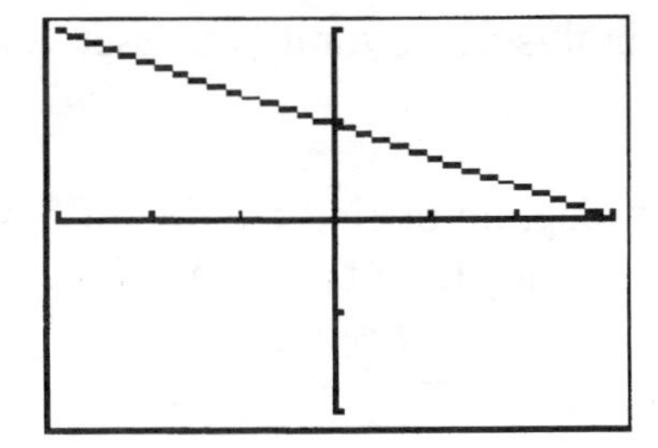

Section 1.4 Scatter Diagrams; Linear Curve Fitting

In Problems 7-14:
(d) Use a graphing utility to draw a scatter diagram.
(e) Use a graphing utility to find the line of best fit.
(f) Use a graphing utility to graph the line of best fit on the scatter diagram.

7.

x	3	4	5	6	7	8	9
y	4	6	7	10	12	14	16

(d) Use a graphing utility to draw a scatter diagram.

In order for our calculator to display the correlation coefficient, we must first set our calculator in the proper mode to display the diagnostics.

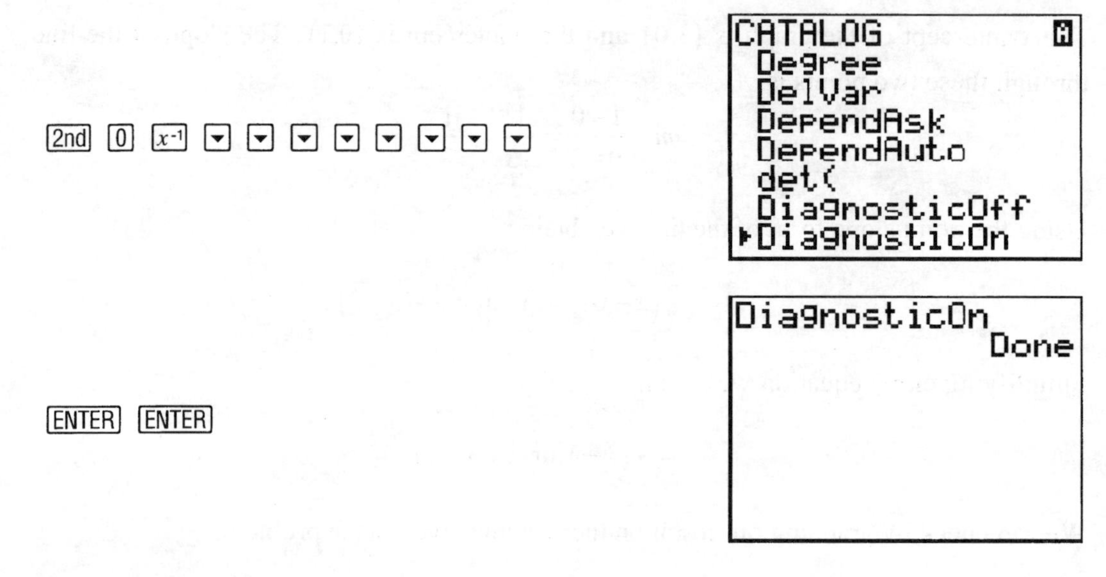

Your calculator will remain in this mode until you reset your calculator or you turn the diagnostics off.

Before we can draw a scatter diagram or find the line of best fit, we must input the data into the calculator. This is done using the `Stat` data editor.

Enter the values for x into L1.

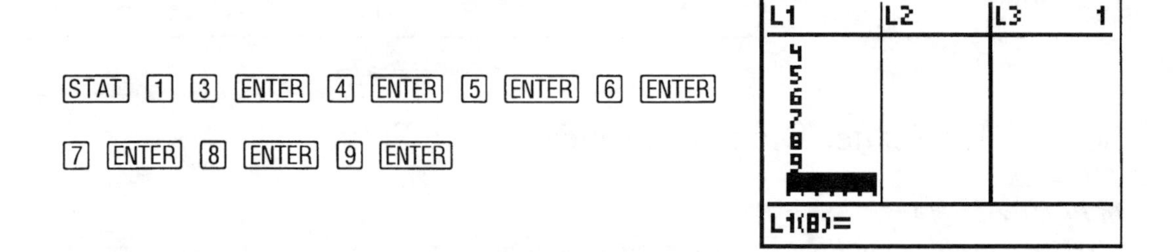

Enter the values for y into L2.

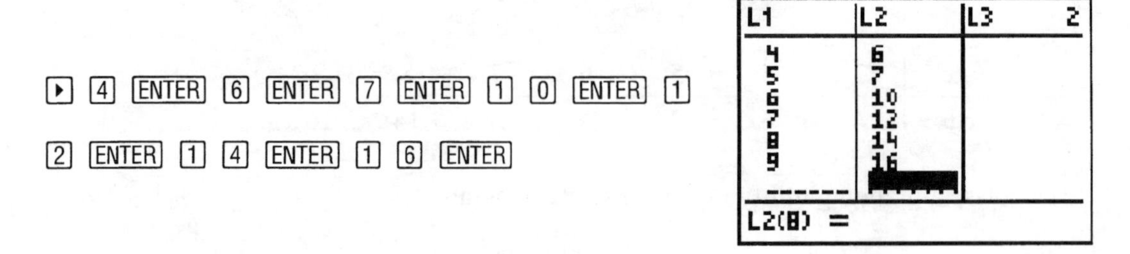

Set an appropriate window for the data. Use the window shown below for this problem.

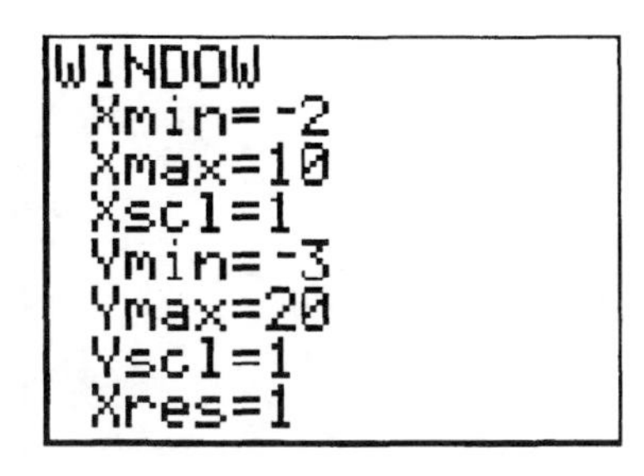

Be sure to clear any functions in the function editor. Go to the [STAT PLOT] to define the scatter plot of the given data. For this example, we will use Plot1 for the scatter diagram.

[2nd] [Y=] [1] [ENTER] [▼] [ENTER] [▼] [2nd] [1] [ENTER] [2nd] [2] [ENTER] [ENTER]

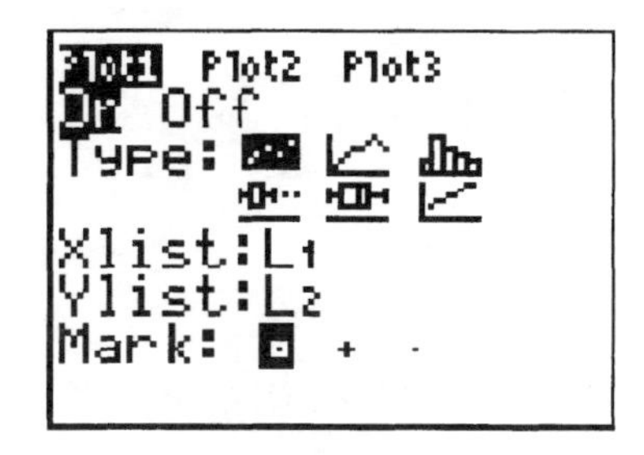

Note that if you have stored the data in lists other than L1 and L2, be sure to designate the correct lists for the *x* and *y* data sets.

Draw the scatter diagram.

[GRAPH]

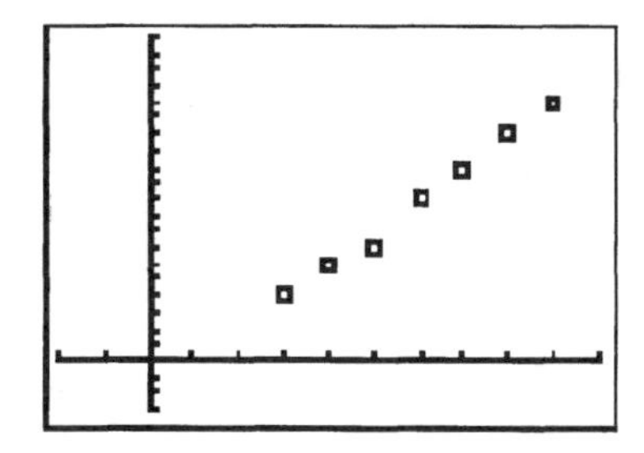

(e) Use a graphing utility to find the line of best fit.

To calculate the line of best fit we will use the LinReg(ax+b) command. The format is

LinReg(ax+b) *xlist*, *ylist*, *yvariable*

where *xlist* is the name of the list of *x* values, *ylist* is the name of the list of *y* values, and *yvariable* is the name of the function in the function editor where you wish to store the line of best fit.

Find the line of best fit and store the expression in y1.

[STAT] [▸] [4] [2nd] [1] [,] [2nd] [2] [,] [VARS] [▸] [1]

[1]

```
LinReg(ax+b) L1,
L2,Y1
```

[ENTER]

```
LinReg
 y=ax+b
 a=2.035714286
 b=-2.357142857
 r²=.9929706601
 r=.9964791318
```

The line of best fit is $y = 2.0357x - 2.3571$.

(f) Use a graphing utility to graph the line of best fit on the scatter diagram.

Graph the line of best fit with the scatter diagram.

[GRAPH]

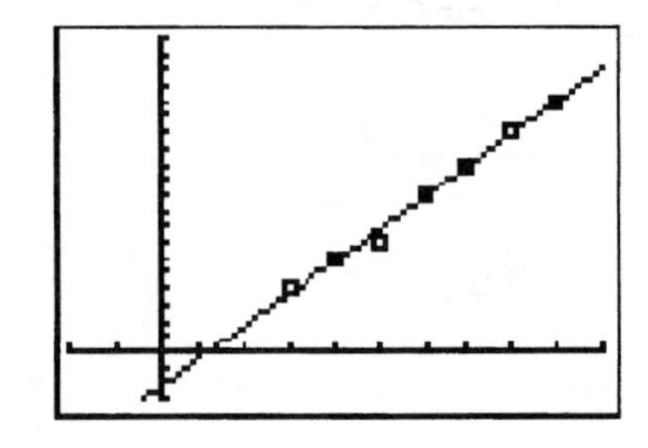

9.

x	–2	–1	0	1	2
y	–4	0	1	4	5

(d) Use a graphing utility to draw a scatter diagram.

Before entering the data for this problems, you will need to clear the lists used in the previous problem.

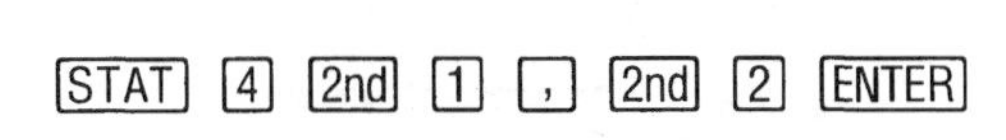

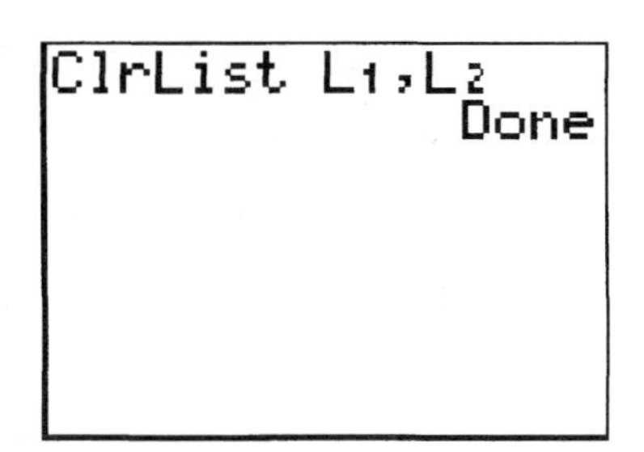

Enter the values for x into L1 and the values for y into L2.

L1	L2	L3 2
-2	-4	
-1	0	
0	1	
1	4	
2	5	

L2(6) =

Define the scatter diagram as Plot1.

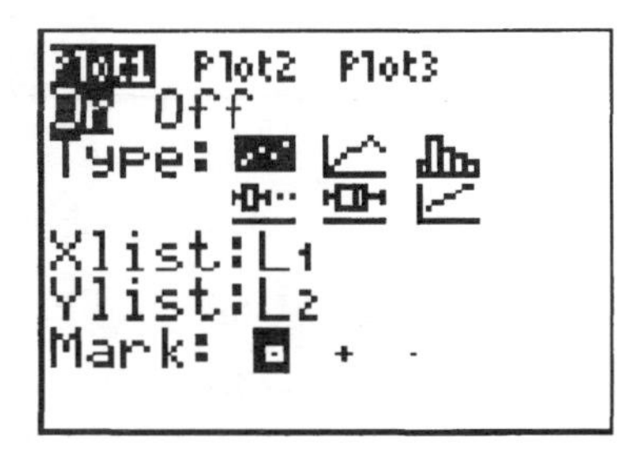

Set an appropriate window and draw the scatter diagram. Be sure to clear any functions in the function editor.

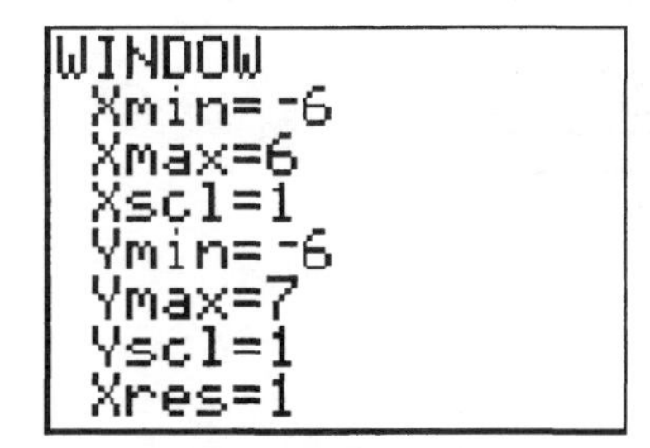

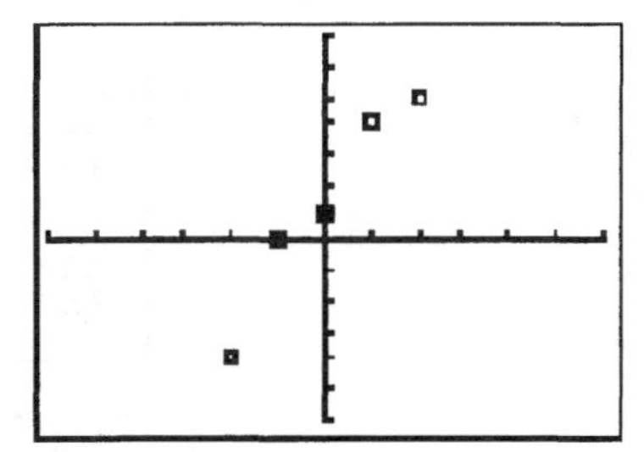

(e) Use a graphing utility to find the line of best fit.

Find the line of best fit and store the equation in y1.

```
LinReg(ax+b) L1,
L2,Y1
```

```
LinReg
 y=ax+b
 a=2.2
 b=1.2
 r²=.9527559055
 r=.9760921604
```

The line of best fit is $y = 2.2x + 1.2$.

(f) Use a graphing utility to graph the line of best fit on the scatter diagram.

Graph the line of best fit with the scatter diagram.

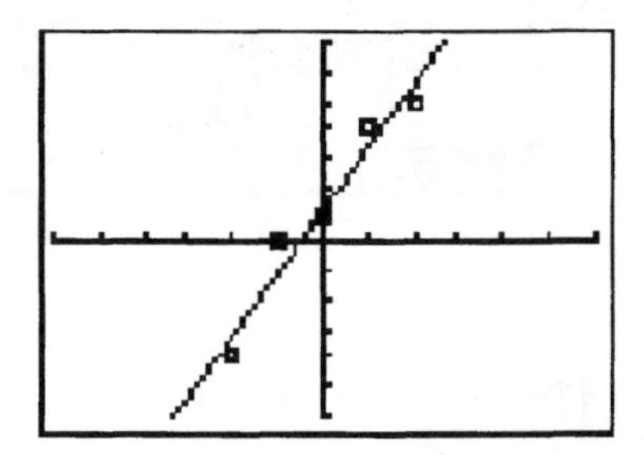

11.

x	20	30	40	50	60
y	100	95	91	83	70

(d) Use a graphing utility to draw a scatter diagram.

Enter the values for x into `L1` and the values for y into `L2`. Be sure to clear the lists first.

L1	L2	L3 2
20	100	------
30	95	
40	91	
50	83	
60	70	

L2(6) =

Define the scatter diagram as `Plot1`.

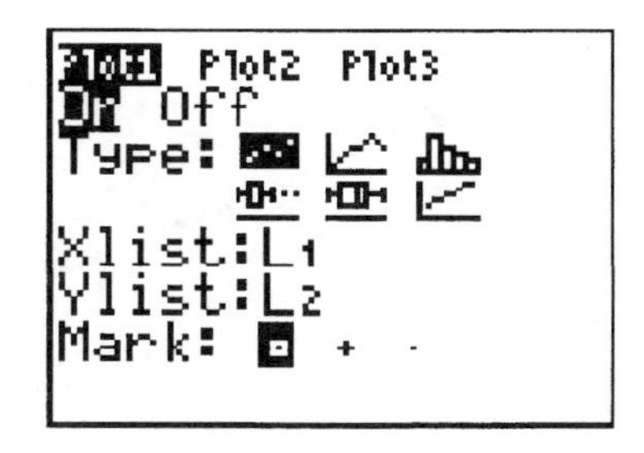

Set an appropriate window and draw the scatter diagram. Be sure to clear any functions in the function editor.

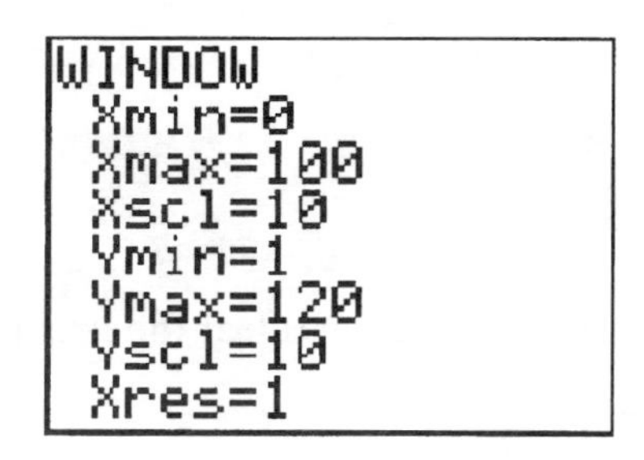

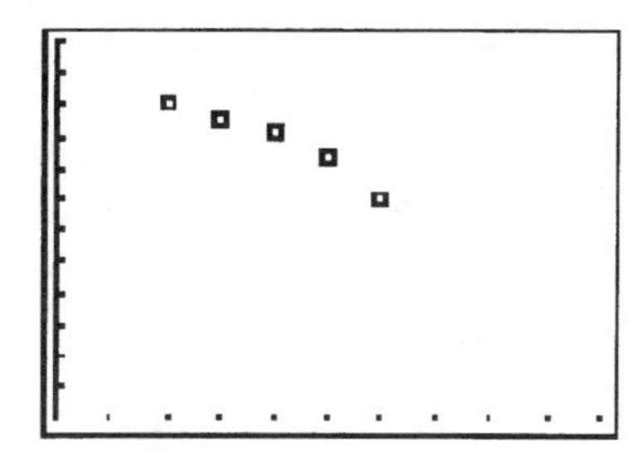

(e) Use a graphing utility to find the line of best fit.

Find the line of best fit and store the equation in y1.

```
LinReg(ax+b) L1,
L2,Y1
```

```
LinReg
 y=ax+b
 a=-.72
 b=116.6
 r²=.9411764706
 r=-.9701425001
```

The line of best fit is $y = -0.72x + 116.6$.

(f) Use a graphing utility to graph the line of best fit on the scatter diagram.

Graph the line of best fit with the scatter diagram.

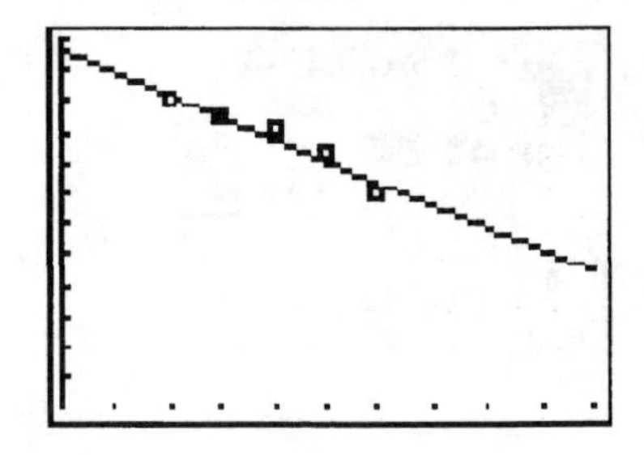

13.

x	−20	−17	−15	−14	−10
y	100	120	118	130	140

(d) Use a graphing utility to draw a scatter diagram.

Enter the values for x into L1 and the values for y into L2. Be sure to clear the lists first.

L1	L2	L3
-20	100	
-17	120	
-15	118	
-14	130	
-10	140	

L2(6) =

Define the scatter diagram as Plot1.

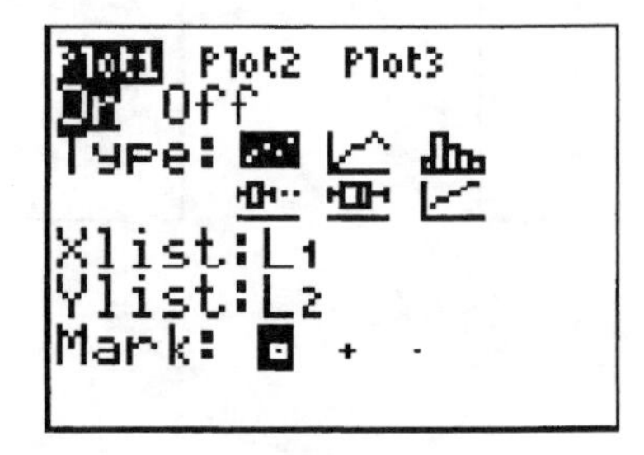

Set an appropriate window and draw the scatter diagram. Be sure to clear any functions in the function editor.

```
WINDOW
 Xmin=-30
 Xmax=10
 Xscl=1
 Ymin=0
 Ymax=160
 Yscl=50
 Xres=1
```

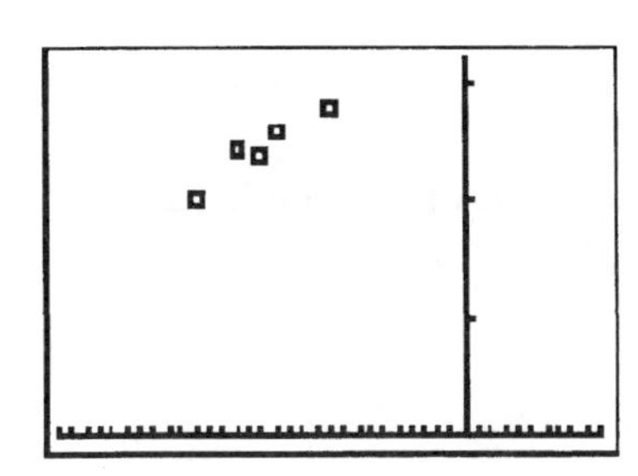

(e) Use a graphing utility to find the line of best fit.

Find the line of best fit and store the equation in y1.

```
LinReg(ax+b) L1,
L2,Y1
```

```
LinReg
 y=ax+b
 a=3.861313869
 b=180.2919708
 r²=.916802081
 r=.957497823
```

The line of best fit is $y = 3.86131x + 180.29197$.

(f) Use a graphing utility to graph the line of best fit on the scatter diagram.

Graph the line of best fit with the scatter diagram.

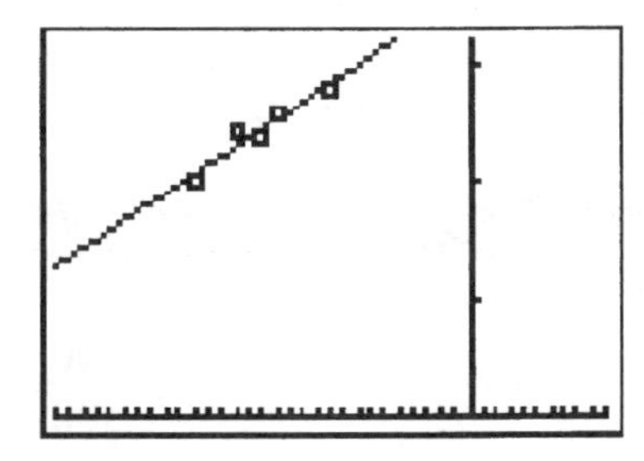

15. **Consumption and Disposable Incomes** An economist wishes to estimate a line that relates personal consumptions expenditures C and disposable income I. Both C and I are in thousands of dollars. She interviews eight heads of households for families of size 3 and obtains the data below. Let I represent the independent variable and C the dependent variable.

***I*(000)**	20	20	18	27	36	37	45	50
***C*(000)**	16	18	13	21	27	26	36	39

(e) Use a graphing utility to find the line of best fit to the data.

Enter the values for I into L1 and the values for C into L2. Be sure to clear the lists first. Find the line of best fit.

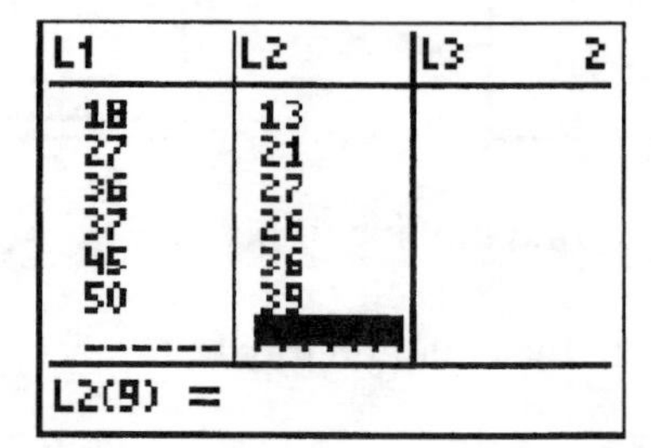


```
LinReg(ax+b) L1,
L2,Y1
```

```
LinReg
 y=ax+b
 a=.7548890222
 b=.6266346731
 r²=.9733118294
 r=.9865656742
```

The line of best fit is $C = 0.75489I + 0.62663$.

17. **Mortgage Calculation** The amount of money that a lending institution will allow you to borrow mainly depends on the interest rate and your annual income. The following data represent the annual income, I, required by a bank in order to lend L dollars at an interest rate of 7.5% for 30 years.

Annual Income, I($)	Loan Amount, L($)
15,000	44,600
20,000	59,500
25,000	74,500
30,000	89,400
35,000	104,300
40,000	119,200
45,000	134,100
50,000	149,000
55,000	163,900
60,000	178,800
65,000	193,700
70,000	208,600

Let I, the annual income in thousands, represent the independent variable and L, the loan amount in thousands, represent the dependent variable.

(a) Use a graphing utility to draw a scatter diagram of the data.

Enter the values for I into L1 and the values for L into L2. Be sure to clear the lists first.

L1	L2	L3 2
45	134.1	
50	149	
55	163.9	
60	178.8	
65	193.7	
70	208.6	

L2(13) =

Define the scatter diagram as Plot1.

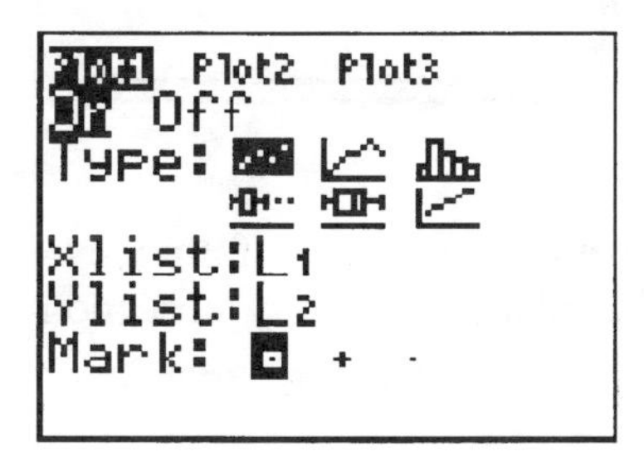

Set an appropriate window and draw the scatter diagram. Be sure to clear any functions in the function editor.

```
WINDOW
 Xmin=0
 Xmax=75.5
 Xscl=5
 Ymin=0
 Ymax=236.5
 Yscl=10
 Xres=1
```

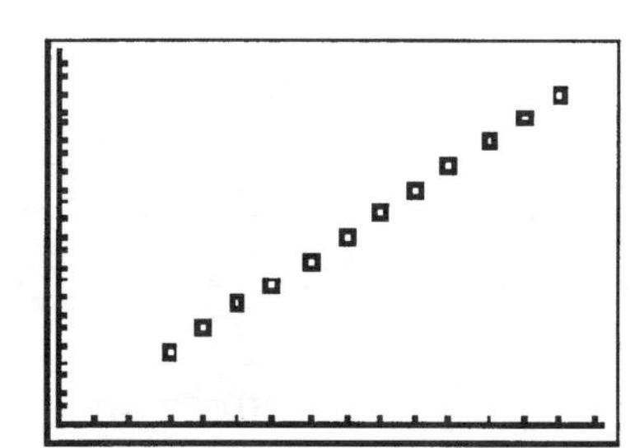

(b) Use a graphing utility to find the line of best fit to the data.

Find the line of best fit and store the equation in y1.

```
LinReg(ax+b) L1,
L2,Y1
```

```
LinReg
 y=ax+b
 a=2.981398601
 b=-.0761072261
 r²=.9999996956
 r=.9999998478
```

The line of best fit is $y = 2.9814x - 0.07611$.

(c) Graph the line of best fit on the scatter diagram drawn in part (a).

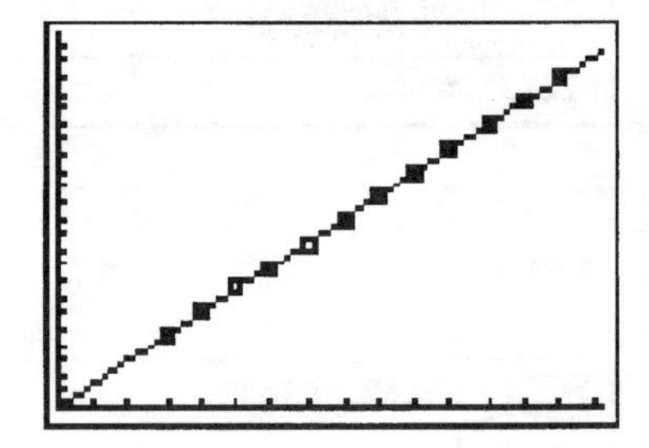

19. **Apparent Room Temperature** The following data represent the apparent temperature versus the relative humidity in a room whose actual temperature is 65° Fahrenheit. Let h represent the independent variable and T the dependent variable.

Relative Humidity, $h(\%)$	Apparent Temperature, $T(\ \mathrm{F})$
0	59
10	60
20	61
30	61
40	62
50	63
60	64
70	65
80	65
90	66
100	67

(a) Use a graphing utility to draw a scatter diagram of the data.

Enter the values for h into L1 and the values for T into L2. Be sure to clear the lists first.

L1	L2	L3 2
50	63	
60	64	
70	65	
80	65	
90	66	
100	67	

L2(12) =

Define the scatter diagram as Plot1.

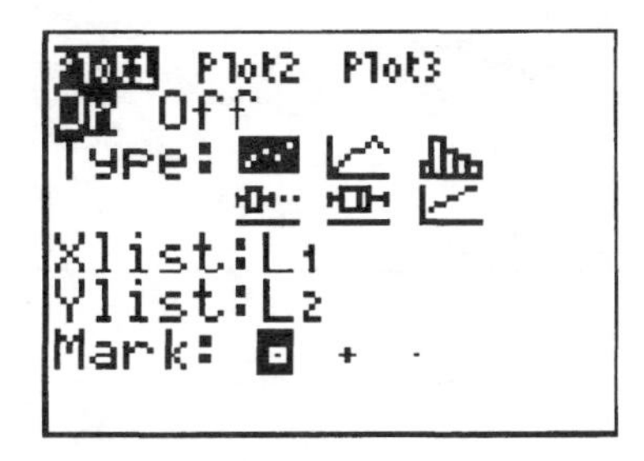

Set an appropriate window and draw the scatter diagram. Be sure to clear any functions in the function editor.

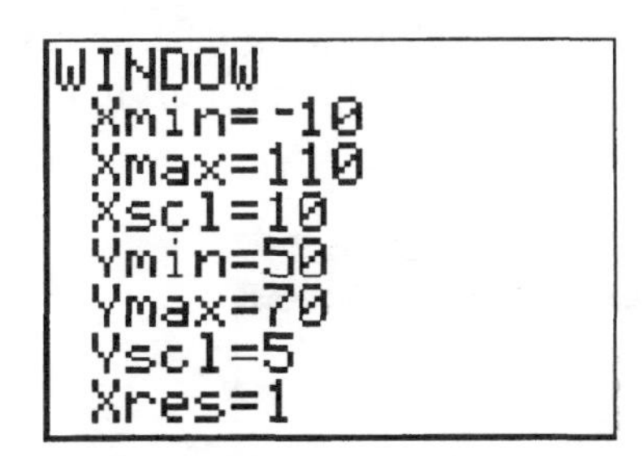

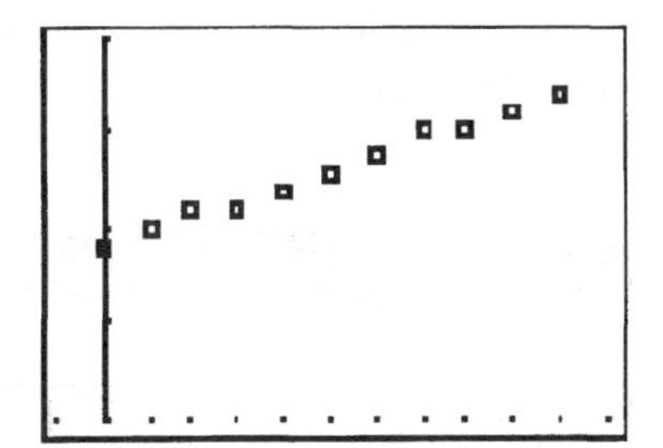

(b) Use a graphing utility to find the line of best fit to the data.

Find the line of best fit and store the equation in y1.

```
LinReg(ax+b) L1,
L2,Y1
```

```
LinReg
 y=ax+b
 a=.0781818182
 b=59.09090909
 r²=.9887700535
 r=.9943691736
```

The line of best fit is $y = 0.07818x + 59.0909$.

(c) Graph the line of best fit on the scatter diagram drawn in part (a).

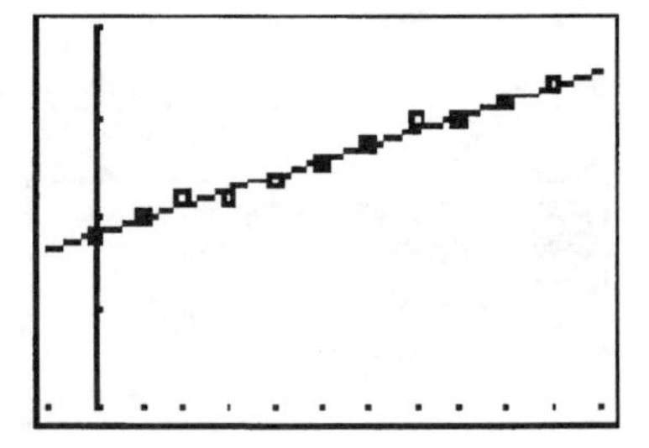

Chapter 1 Review

45. **Concentration of Carbon Monoxide in the Air** The following data represent the average concentration of carbon monoxide in parts per million (ppm) in the air for 1987 – 1993.

Year	Concentration of Carbon Monoxide (ppm)
1987	6.69
1988	6.38
1989	6.34
1990	5.87
1991	5.55
1992	5.18
1993	4.88

(f) Use a graphing utility to find the slope of the line of best fit for these data.

Enter the values for the year into L1 and the values for the average level of carbon monoxide into L2. Be sure to clear the lists first.

L1	L2	L3
1988	6.38	
1989	6.34	
1990	5.87	
1991	5.55	
1992	5.18	
1993	4.88	

L2(8) =

Find the line of best fit and store the equation in y1.

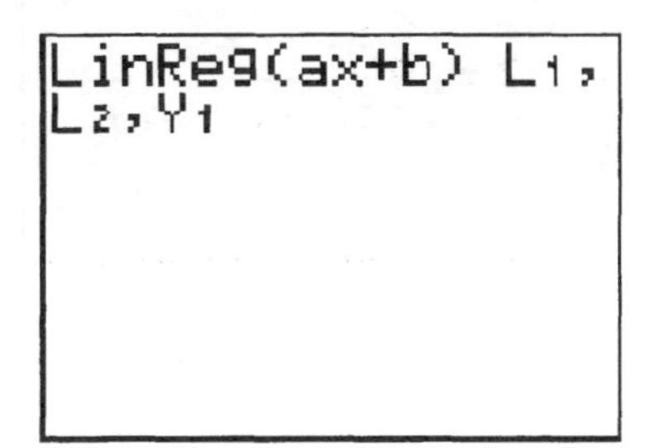

LinReg
y=ax+b
a=-.3078571429
b=618.4771429
r²=.980579603
r=-.9902421941

The slope of the line of best fit is -0.308.

(g) Interpret this slope.

The concentration of carbon monoxide is decreasing by 0.308 ppm per year.

47. **Value of a Portfolio** The following data represent the value of the Vanguard 500 Index Fund for 1996 – 1999.

Year	Value per Share
1996	$ 69.17
1997	90.07
1998	113.95
1999	135.33

(e) Use a graphing utility to find the line of best fit for these data.

Enter the values for the year into L1 and the values for the value per share into L2. Be sure to clear the lists first.

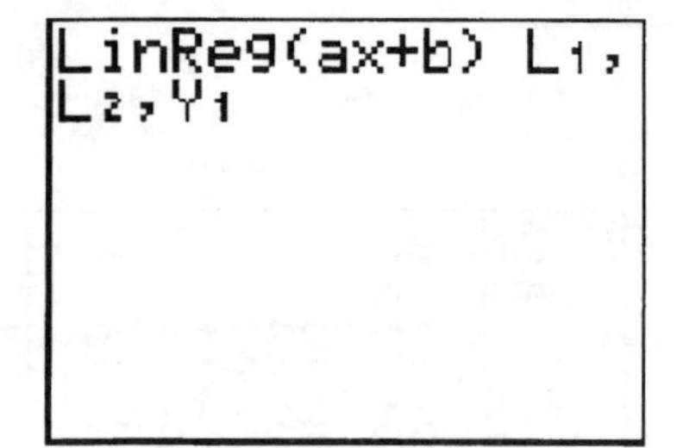

Find the line of best fit and store the equation in y1.

```
LinReg(ax+b) L1,
L2,Y1
```

```
LinReg
 y=ax+b
 a=22.236
 b=-44314.28
 r²=.9993697361
 r=.9996848184
```

The line of best fit is $y = 22.236x - 44314.28$.

Summary

The command introduced in this chapter is:

`LinReg(ax+b)` *xlist*, *ylist*, *yvariable*

Chapter 2 – Systems of Linear Equations; Matrices

Section 2.2 Systems of Linear Equations: Matrix Method

In problems 59–64, use a graphing utility to find the row-echelon form (REF) and reduced row-echelon form (RREF) of the augmented matrix of each of the following systems. Solve each system. If the system has no solution, say it is inconsistent.

59. $$\begin{cases} 2x - 2y + z = 2 \\ x - \frac{1}{2}y + 2z = 1 \\ 2x + \frac{1}{3}y - z = 0 \end{cases}$$

Before we can find either form of the matrix, we must first enter the system into a matrix using the matrix editor.

Enter the system into the matrix [A].

[2nd] [x^{-1}] [◄] [1] [3] [ENTER] [4] [ENTER]

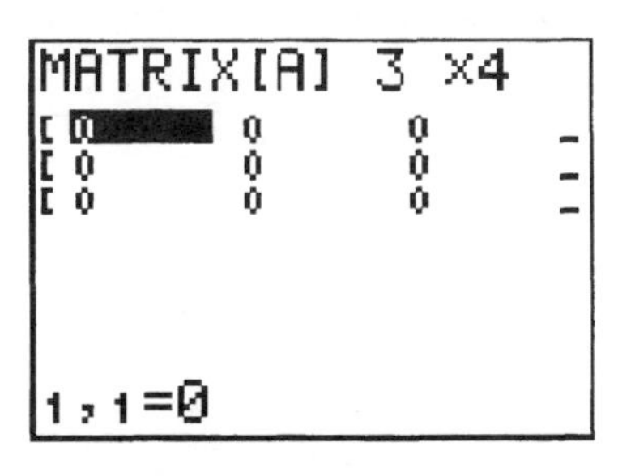

[2] [ENTER] [(-)] [2] [ENTER] [1] [ENTER] [2] [ENTER] [1] [ENTER] [(-)] [1] [÷] [2] [ENTER] [2] [ENTER] [1] [ENTER] [2] [ENTER] [1] [÷] [3] [ENTER] [(-)] [1] [ENTER] [0] [ENTER]

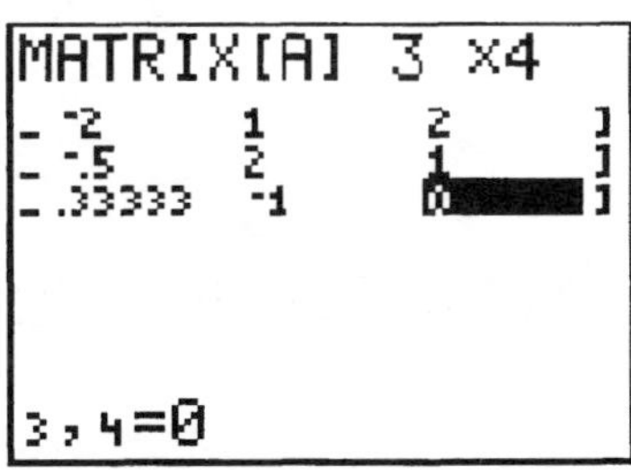

When you are done, press [2nd] [MODE] to quit the matrix editor.

Both the `ref` and `rref` commands can be found under `MATH` submenu of the `[MATRX]` menu. The format for the commands are:

`ref`(*matrixname*)

`rref`(*matrixname*)

Find the row-echelon form of the matrix `[A]`.

[2nd] [x^{-1}] [▶] [▲] [▲] [▲] [▲] [▲] [▲]

```
NAMES MATH EDIT
0↑cumSum(
A:ref(
B:rref(
C:rowSwap(
D:row+(
E:*row(
F:*row+(
```

[ENTER] [2nd] [x^{-1}] [1] [)] [ENTER]

```
ref([A])
[[1  -1  .5           ...
 [0  1   -.857142...
 [0  0   1            ...
```

To see the rest of the matrix, use the [▶] key to scroll through and view the other entries.

[▶] [▶] [▶] [▶] [▶] [▶]

```
ref([A])
....5               1...
...-.8571428571     -...
...1                ....
```

[▶] [▶] [▶] [▶] [▶] [▶] [▶] [▶] [▶] [▶] [▶] [▶]

```
ref([A])
...  1              ]
...  -.8571428571]
...  .2222222222  ]]
```

You can also have the calculator display the decimal results in fraction form (when possible). You can do this by using the ▷Frac command. The format is

▷Frac *expression*

If possible, display the decimal results in fraction form.

[2nd] [(-)] [MATH] [1] [ENTER]

```
Ans▸Frac
[[1  -1  1/2   1   ...
 [0  1   -6/7  -6/...
 [0  0   1     2/9...
```

[▸] [▸]

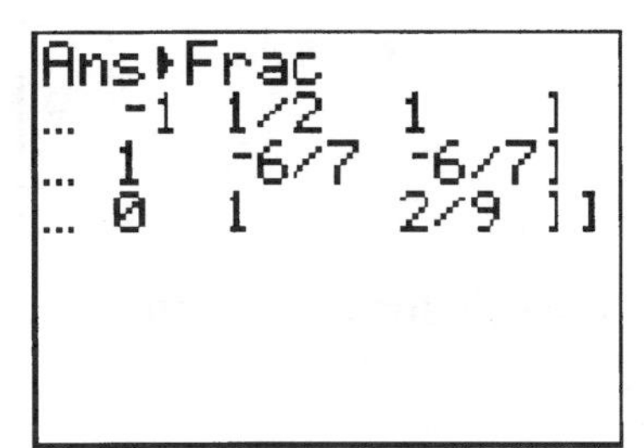

Find the reduced row-echelon form of the matrix [A] and display decimal numbers in fraction form (if possible).

[2nd] [x^{-1}] [▸] [▴] [▴] [▴] [▴] [▴] [ENTER] [2nd] [x^{-1}] [1] [)]

[MATH] [1] [ENTER]

```
rref([A])▸Frac
    [[1 0 0 2/9 ]
     [0 1 0 -2/3]
     [0 0 1 2/9 ]]
```

The solution to the system is $x=\frac{2}{9}$, $y=-\frac{2}{3}$, and $z=\frac{2}{9}$.

61. $\begin{cases} x+y+z= & 4 \\ x-y-z= & 0 \\ \quad y-z= & -4 \end{cases}$

Enter the system into the matrix [A]. Find the row-echelon form of the matrix and display the decimal results in fraction from (if possible).

MATRIX[A] 3 ×4

3,4=-4

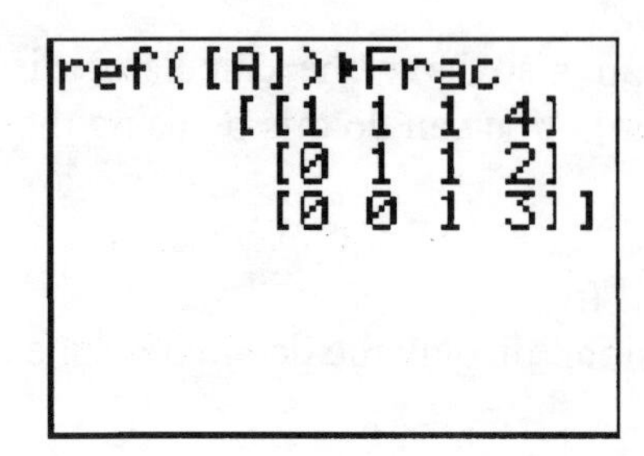

Find the reduced row-echelon form of the matrix [A] and display decimal numbers in fraction form (if possible).

rref([A])▶Frac
[[1 0 0 2]
[0 1 0 -1]
[0 0 1 3]]

The solution to the system is $x = 2$, $y = -1$, and $z = 3$.

63. $$\begin{cases} x_1 + x_2 + x_3 + \ \ x_4 = 20 \\ \qquad x_2 + x_3 + \ \ x_4 = \ \ 0 \\ \qquad\qquad x_3 + \ \ x_4 = 13 \\ \qquad x_2 \qquad - 2x_4 = -5 \end{cases}$$

Enter the system into the matrix [A]. Find the row-echelon form of the matrix and display the decimal results in fraction from (if possible).

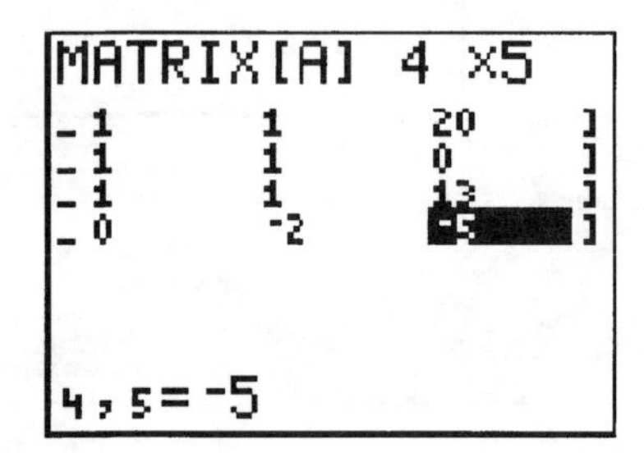

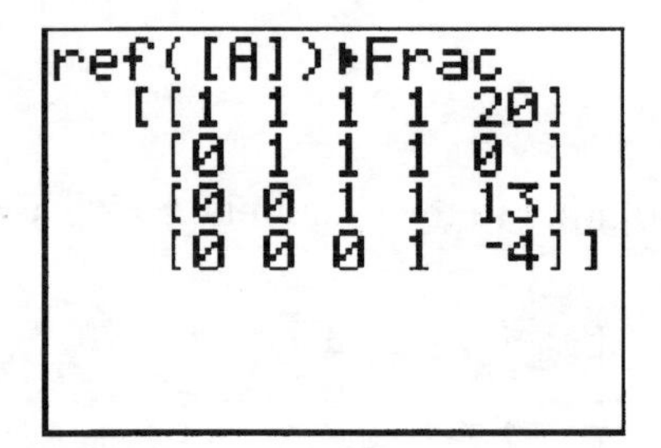

Find the reduced row-echelon form of the matrix [A] and display decimal numbers in fraction form (if possible).

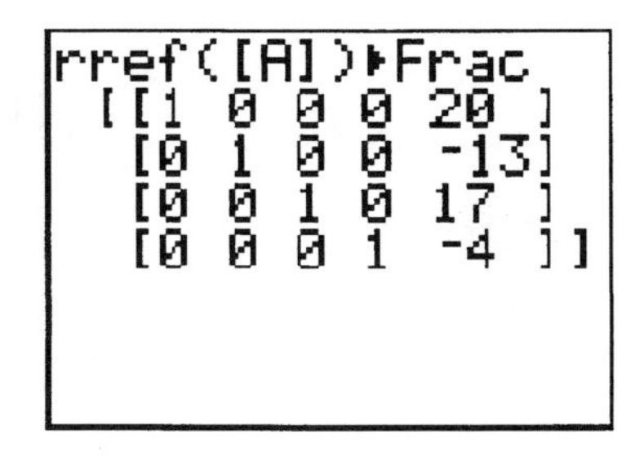

The solution to the system is $x_1 = 20$, $x_2 = -13$, $x_3 = 17$, and $x_4 = -4$.

Section 2.4 Matrix Algebra

In Problems 57–62, use a graphing utility to perform the indicated operations on the matrices given below.

$$A = \begin{bmatrix} -1 & -1 & 3 & 0 \\ 2 & 6 & 2 & 2 \\ -4 & 2 & 3 & 2 \\ 7 & 0 & 5 & -1 \end{bmatrix} \quad B = \begin{bmatrix} -1 & 2 & 4 & 5 \\ 2 & 0 & 5 & 3 \\ 0.5 & 6 & -7 & 11 \\ 5 & -1 & 2 & 7 \end{bmatrix} \quad C = \begin{bmatrix} 13 & -8 & 7 & 0 \\ 0 & 5 & 0 & -2 \\ 5 & 0 & 7 & 0 \\ 7 & 7 & 7 & 7 \end{bmatrix}$$

57. $A + B$

First, enter the matrices A, and B into `[A]`, and `[B]`, respectively, on your calculator.

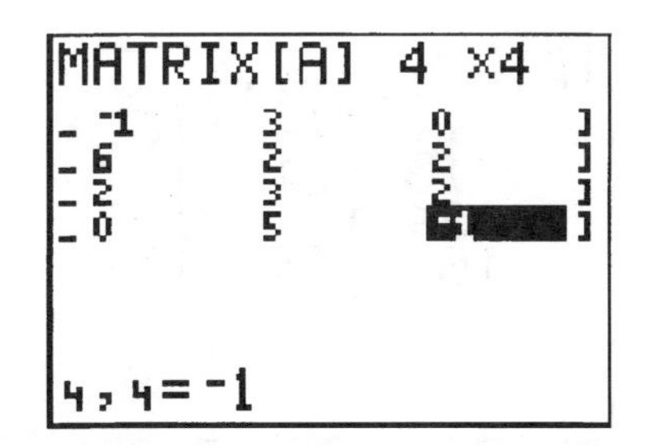

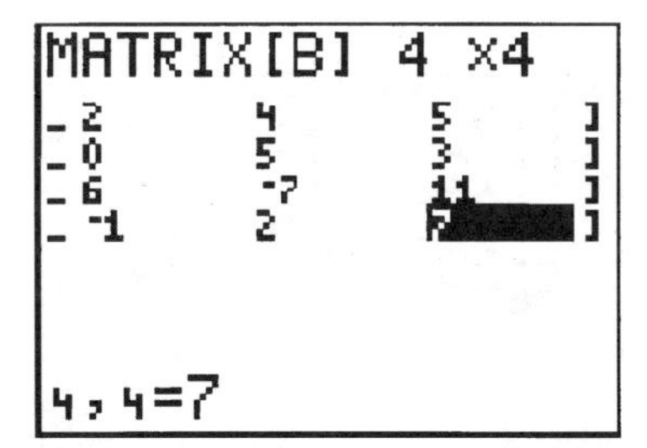

To find the sum or difference of two matrices, use the [+] or [−] key. To enter the name of a matrix, go to the `NAMES` submenu of `[MATRX]` and select the desired name.

Find the sum $A + B$.

```
[A]+[B]
[[-2    1   7  5 ...
 [4     6   7  5 ...
 [-3.5  8  -4  13...
 [12   -1   7  6 ...
```

2nd x^{-1} 1 + 2nd x^{-1} 2 ENTER

Thus, $A+B=\begin{bmatrix} -2 & 1 & 7 & 5 \\ 4 & 6 & 7 & 5 \\ -3.5 & 8 & -4 & 13 \\ 12 & -1 & 7 & 6 \end{bmatrix}$

59. $C-3(A+B)$

First, enter the matrices A, B, and C into [A], [B], and [C], respectively, on your calculator. If you did problem 57, then you will only need to enter the matrix C into [C].

```
MATRIX[C] 4 ×4
- -8    7    0     ]
- 5     0    -2    ]
- 0     7    0     ]
- 7     7    7     ]

4,4=7
```

Find $C-3(A+B)$.

```
[C]-3([A]+[B])
[[19    -11  -14 ...
 [-12   -13  -21 ...
 [15.5  -24  19  ...
 [-29   10   -14 ...
```

```
[C]-3([A]+[B])
...   -11  -14  -15]
...   -13  -21  -17]
...5  -24  19   -39]
...   10   -14  -11]]
```

$$\text{Thus, } C-3(A+B)=\begin{bmatrix} 19 & -11 & -14 & -15 \\ -12 & -13 & -21 & -17 \\ 15.5 & -24 & 19 & -39 \\ -29 & 10 & -14 & -11 \end{bmatrix}$$

61. $3(B+C)-A$

If they are not already stored in your calculator, enter the matrices A, B, and C into `[A]`, `[B]`, and `[C]`, respectively, on your calculator.

Find $3(B+C)-A$.

```
3([B]+[C])-[A]
[[37    -17 30 1...
 [4     9   13 1...
 [20.5 16   -3 3...
 [29   18   22 4...
```

```
3([B]+[C])-[A]
...7    -17 30 15]
...     9   13 1 ]
...0.5 16   -3 31]
...9   18   22 43]]
```

$$\text{Thus, } 3(B+C)-A=\begin{bmatrix} 37 & -17 & 30 & 15 \\ 4 & 9 & 13 & 1 \\ 20.5 & 16 & -3 & 31 \\ 29 & 18 & 22 & 43 \end{bmatrix}$$

Section 2.5 Multiplication of Matrices

In Problems 43–50, use a graphing utility to perform the indicated operations on the matrices given below.

$$A=\begin{bmatrix} -1 & -1 & 3 & 0 \\ 2 & 6 & 2 & 2 \\ -4 & 2 & 3 & 2 \\ 7 & 0 & 5 & -1 \end{bmatrix} \quad B=\begin{bmatrix} -1 & 2 & 4 & 5 \\ 2 & 0 & 5 & 3 \\ 0.5 & 6 & -7 & 11 \\ 5 & -1 & 2 & 7 \end{bmatrix} \quad C=\begin{bmatrix} 13 & -8 & 7 & 0 \\ 0 & 5 & 0 & -2 \\ 5 & 0 & 7 & 0 \\ 7 & 7 & 7 & 7 \end{bmatrix}$$

43. *AB*

First, enter the matrices *A*, and *B* into [A], and [B], respectively, on your calculator.

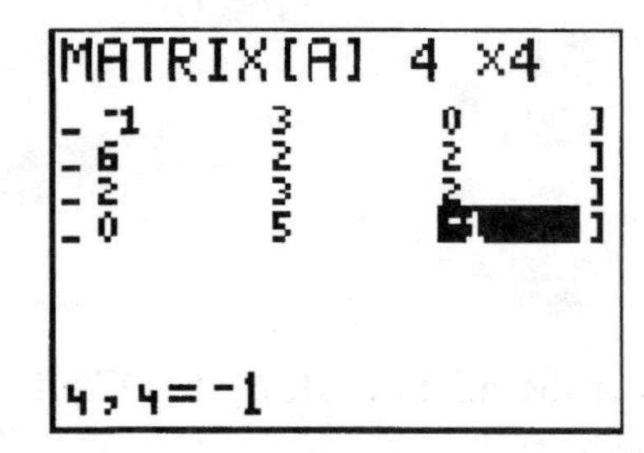

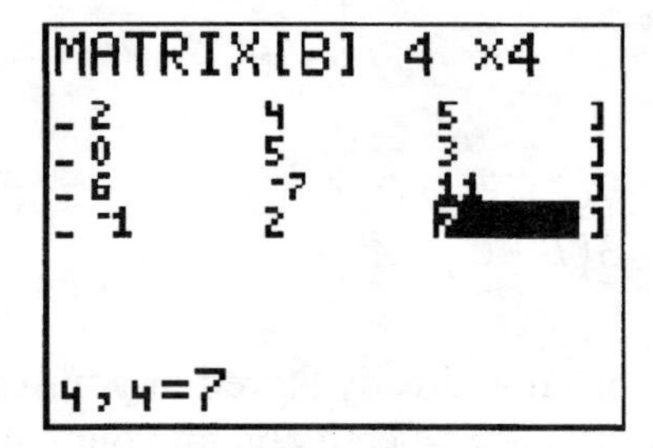

To find the product of two matrices, use the [×] key. To enter the name of a matrix, go to the NAMES submenu of [MATRX] and select the desired name.

Find the product AB.

[2nd] [x^{-1}] [1] [×] [2nd] [x^{-1}] [2] [ENTER]

```
[A]*[B]
[[.5    16  -30  2...
 [21    14  28   6...
 [19.5  8   -23  3...
 [-9.5  45  -9   8...
```

Scroll through the rest of the result.

```
[A]*[B]
...5    16  -30  25]
...1    14  28   64]
...9.5  8   -23  33]
...9.5  45  -9   83]]
```

$$\text{Thus, } AB = \begin{bmatrix} 0.5 & 16 & -30 & 25 \\ 21 & 14 & 28 & 64 \\ 19.5 & 8 & -23 & 33 \\ -9.5 & 45 & -9 & 83 \end{bmatrix}$$

45. $(AB)C$

First, enter the matrices A, B, and C into [A], [B], and [C], respectively, on your calculator. If you did problem 43, then you will only need to enter the matrix C into [C].

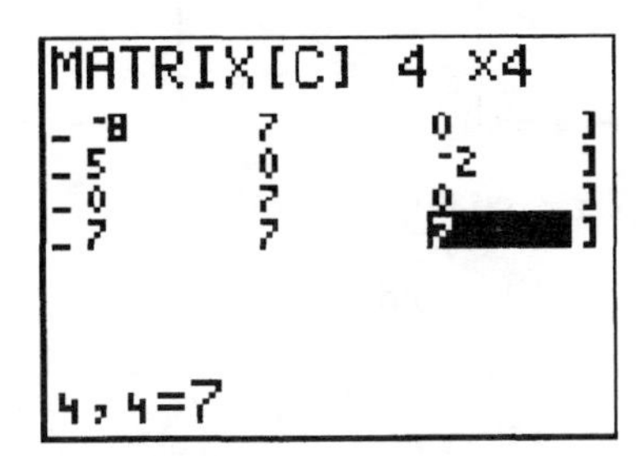

Find $(AB)C$.

```
([A]*[B])*[C]
[[31.5   251  -31...
 [861    350  791...
 [369.5 115  206...
 [412.5 882  451...
```

```
([A]*[B])*[C]
...251  -31.5 143]
...350 791    420]
...115 206.5 215]
...882 451.5 491]]
```

Thus, $(AB)C = \begin{bmatrix} 31.5 & 251 & -31.5 & 143 \\ 861 & 350 & 791 & 420 \\ 369.5 & 115 & 206.5 & 215 \\ 412.5 & 882 & 451.5 & 491 \end{bmatrix}$

47. $B(A+C)$

If they are not already stored in your calculator, enter the matrices A, B, and C into [A], [B], and [C], respectively, on your calculator.

Find $B(A+C)$. Be sure t use the ☒ for multiplication.

```
[B]*([A]+[C])
[[66  74     94 ...
 [71  13     106...
 [165 124.5 79 ...
 [158 -3     152...
```

```
[B]*([A]+[C])
... 74     94   38]
... 13     106  28]
... 124.5  79   52]
... -3     152  46]]
```

Thus, $B(A+C)=\begin{bmatrix} 66 & 74 & 94 & 38 \\ 71 & 13 & 106 & 28 \\ 165 & 124.5 & 79 & 52 \\ 158 & -3 & 152 & 46 \end{bmatrix}$

49. $A(2B-3C)$

If they are not already stored in your calculator, enter the matrices A, B, and C into [A], [B], and [C], respectively, on your calculator.

Find $A(2B-3C)$. Be sure to use the ⊠ for multiplication.

```
[A]*(2[B]-3[C])
[[-5   23    -10...
 [-108 -56   -70...
 [108  -152  -67...
 [-346 279   -24...
```

```
[A]*(2[B]-3[C])
...23    -102 44 ]
...-56   -70  122]
...-152  -67  36 ]
...279   -249 187]]
```

Thus, $A(2B-3C)=\begin{bmatrix} -5 & 23 & -102 & 44 \\ -108 & -56 & -70 & 122 \\ 108 & -152 & -67 & 36 \\ -346 & 279 & -249 & 187 \end{bmatrix}$

In Problems 67–69, use a graphing utility to compute A^2, A^{10}, and A^{15}.

67. $A = \begin{bmatrix} -0.5 & -1 & 0.3 & 0 & 0.3 \\ 2 & 1.6 & 1 & -1 & 0.4 \\ -4 & 2 & 0.7 & 2 & 0.2 \\ 1 & 0 & 0 & -1 & 0 \\ 0 & 0 & -0.9 & 0 & 0 \end{bmatrix}$

First, enter the matrix A into [A] on your calculator.

Raising a matrix to a power on your calculator is no different than raising a number to a power on your calculator.

Find A^2.

[2nd] [x⁻¹] [1] [^] [2] [ENTER]

```
[A]^2
[[-2.95  -.5    -1...
 [-2.8   2.56   2....
 [5.2    8.6    1....
 [-1.5   -1     .3...
 [3.6    -1.8   -....
```

Scroll through the rest of the matrix.

```
[A]^2
...5   -1.21  1.6   ...
...56  2.54   1.4   ...
...6   1.11   -2.6  ...
...    .3     1     ...
....8  -.63   -1.8  ...
```

```
[A]^2
....21  1.6   -.49]
...54   1.4   1.44]
...11   -2.6  -.26]
...     1     .3  ]
...63   -1.8  -.18]]
```

Thus, $A^2 = \begin{bmatrix} -2.95 & -0.5 & -1.21 & 1.6 & -0.49 \\ -2.8 & 2.56 & 2.54 & 1.4 & 1.44 \\ 5.2 & 8.6 & 1.11 & -2.6 & -0.26 \\ -1.5 & -1 & 0.3 & 1 & 0.3 \\ 3.6 & -1.8 & -0.63 & -1.8 & -0.18 \end{bmatrix}$.

Find A^{10}.

```
[A]^10
[[433.0971024  -...
 [1207.69485   5...
 [-1423.02277  8...
 [479.4763646  -...
 [-899.340526  -...
```

Notice our entries have up to seven places after the decimal. We can have the calculator round our answer to at most nine places after the decimal. You can make this change in the [MODE] menu. The default setting is `Float`.

Set the calculator to round the result to three places after the decimal.

[MODE] [▼] [▶] [▶] [▶] [▶] [ENTER]

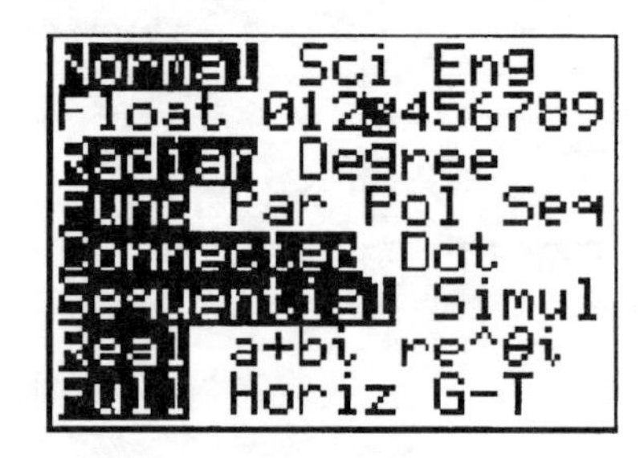

Find A^{10}, round your entries to three places after the decimal.

```
[A]^10
[[433.097      -15...
 [1207.695     599...
 [-1423.023    806...
 [479.476      -24...
 [-899.341     -30...
```

```
[A]^10
...    -1583.562  -3...
...     5998.438  25...
...3    8065.070  42...
...    -246.538   -2...
...    -3064.313  -1...
```

```
[A]^10
...562  -369.817   ...
...38   2563.227   ...
...70   4271.798   ...
...38   -231.543   ...
...313  -1627.502  ...
```

```
[A]^10
...7    -216.548  -1...
...7    -603.847  10...
...8    711.511   20...
...3    -239.737  -1...
...02   449.670   -6...
```

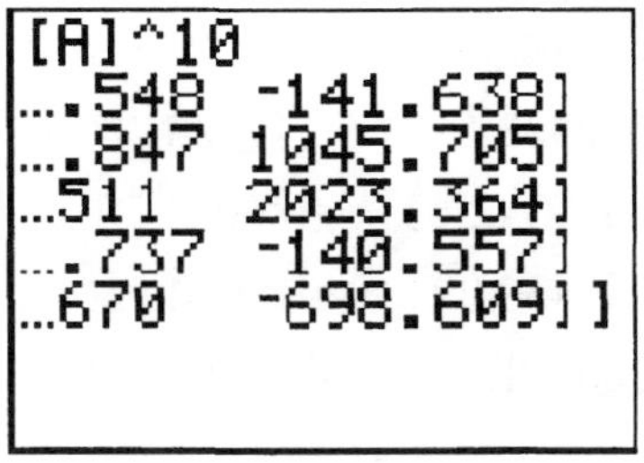

```
[A]^10
....548  -141.638]
....847  1045.705]
...511   2023.364]
....737  -140.557]
...670   -698.609]]
```

Thus $A^{10} = \begin{bmatrix} 433.097 & -1583.562 & -369.817 & -216.548 & -141.638 \\ 1207.695 & 5998.438 & 2563.227 & -603.847 & 1045.705 \\ -1423.023 & 8065.070 & 4271.798 & 711.511 & 2023.364 \\ 479.476 & -246.538 & -231.543 & -239.737 & -140.557 \\ -899.341 & -3064.313 & -1627.502 & 449.670 & -698.609 \end{bmatrix}$.

Find A^{15}, round your entries to three places after the decimal.

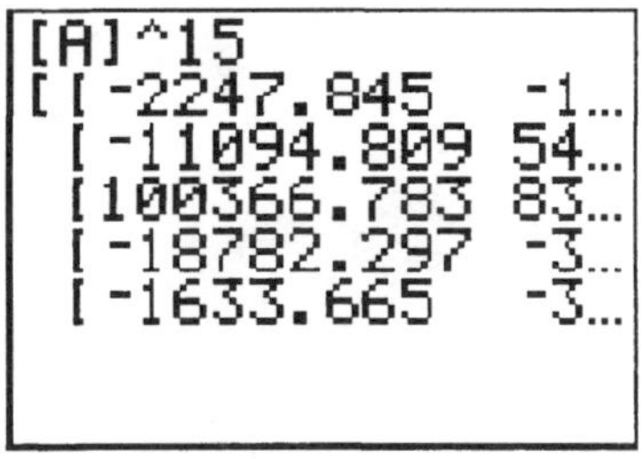

```
[A]^15
[[-2247.845   -1...
 [-11094.809 54...
 [100366.783 83...
 [-18782.297  -3...
 [-1633.665   -3...
```

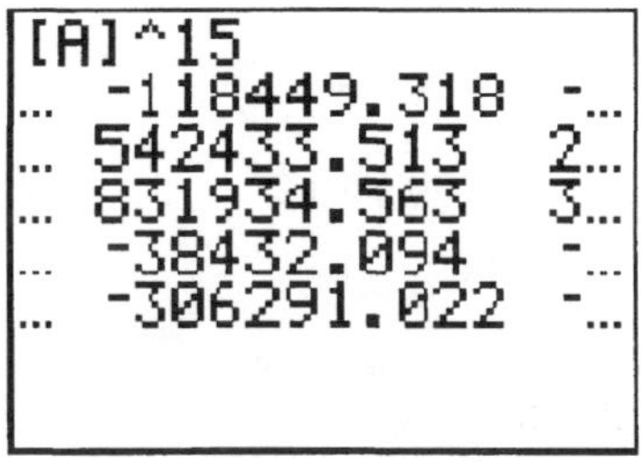

```
[A]^15
...  -118449.318  -...
...  542433.513   2...
...  831934.563   3...
...  -38432.094   -...
...  -306291.022  -...
```

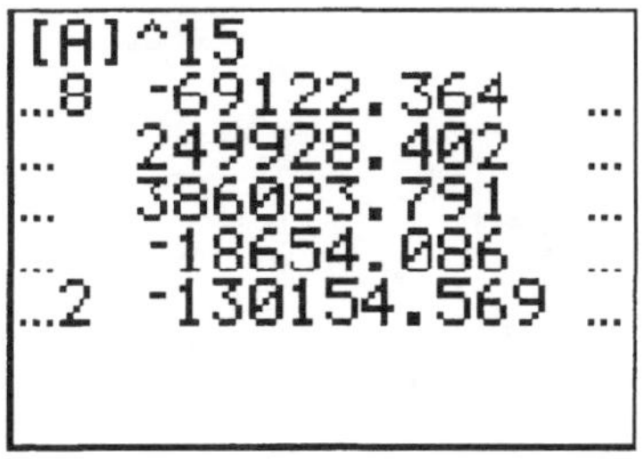

```
[A]^15
...8  -69122.364   ...
...   249928.402   ...
...   386083.791   ...
...   -18654.086   ...
...2  -130154.569  ...
```

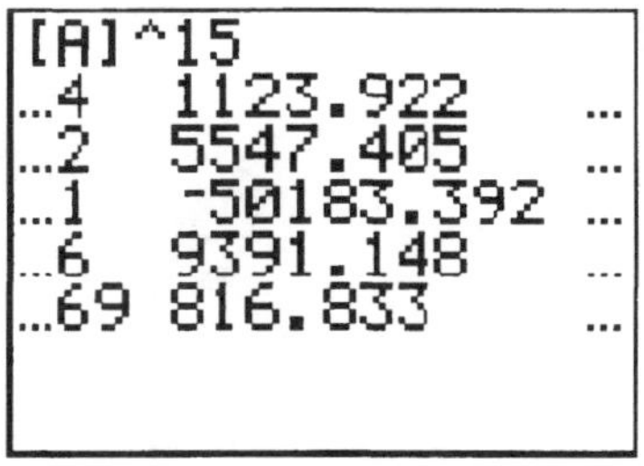

```
[A]^15
...4  1123.922    ...
...2  5547.405    ...
...1  -50183.392  ...
...6  9391.148    ...
...69 816.833     ...
```

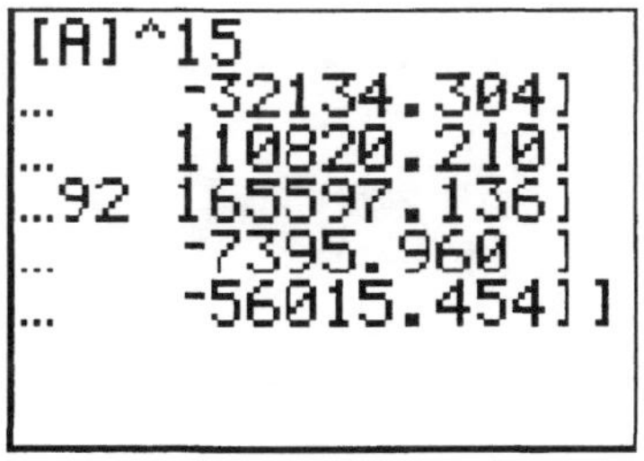

```
[A]^15
...     -32134.304]
...     110820.210]
...92   165597.136]
...     -7395.960 ]
...     -56015.454]]
```

Thus $A^{15} = \begin{bmatrix} -2247.845 & -118449.318 & -69122.364 & 1123.922 & -32134.304 \\ -11094.809 & 542433.513 & 249928.402 & 5547.405 & 110820.210 \\ 100366.783 & 831934.563 & 386083.791 & -50183.392 & 165597.136 \\ -18782.297 & -38432.094 & -18654.086 & 9391.148 & -7395.960 \\ -1633.665 & -306291.022 & -130154.569 & 816.833 & -56015.454 \end{bmatrix}$.

To return the calculator to `Float`, go back to [MODE] and select `Float`.

[MODE] [▼] [ENTER]

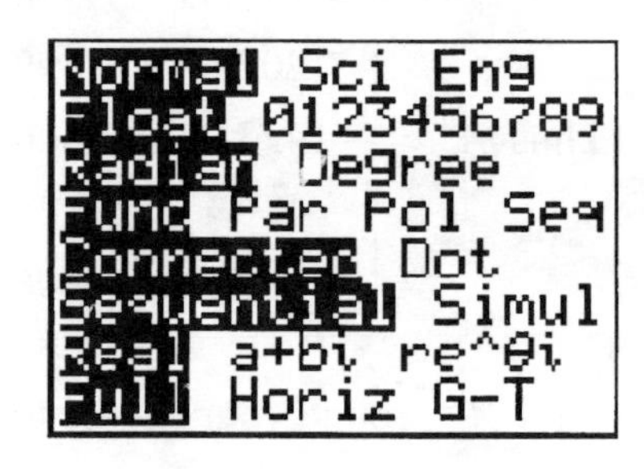

69. $A = \begin{bmatrix} 0 & 1 & 0 \\ 1 & 0 & 1 \\ 1 & 1 & 1 \end{bmatrix}$

First, enter the matrix A into `[A]` on your calculator.

Find A^2.

```
[A]^2
          [[1 0 1]
           [1 2 1]
           [2 2 2]]
```

Thus $A^2 = \begin{bmatrix} 1 & 0 & 1 \\ 1 & 2 & 1 \\ 2 & 2 & 2 \end{bmatrix}$.

Find A^{10}.

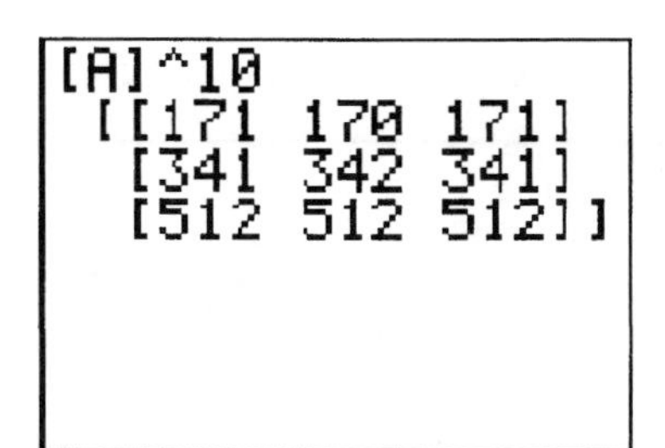

Thus $A^{10} = \begin{bmatrix} 171 & 170 & 171 \\ 341 & 342 & 341 \\ 512 & 512 & 512 \end{bmatrix}$.

Find A^{15}.

```
[A]^15
[[5461  5462  5...
 [10923 10922 1...
 [16384 16384 1...
```

```
[A]^15
...   5462  5461 ]
...3 10922 10923]
...4 16384 16384]]
```

Thus $A^{15} = \begin{bmatrix} 5461 & 5462 & 5461 \\ 10923 & 10922 & 10923 \\ 16384 & 16384 & 16384 \end{bmatrix}$.

Section 2.6 The Inverse of a Matrix

In Problems 51–56, use a graphing utility or EXCEL to find the inverse, if it exists, of each matrix.

51. $\begin{bmatrix} 25 & 61 & -12 \\ 18 & -2 & 4 \\ 8 & 35 & 21 \end{bmatrix}$

First, enter the matrix A into `[A]` on your calculator.

You can find the inverse of a matrix (if it exists) using the [x^{-1}] key. The format is

matrixname^{-1}

Find the inverse of [A].

[2nd] [x^{-1}] [1] [x^{-1}] [ENTER]

```
[A]-1
[[.0054477969   ...
 [.0103568008   ...
 [-.0193366858  ...
```

Scroll through the rest of the inverse.

```
[A]-1
... .0509159483   ...
... -.0185883621  ...
... .0115840517   ...
```

```
[A]-1
...    -.006585249]
...1 .0094588123]
...   .0343630268]]
```

Rounding the entries to four places after the decimal we obtain

$$A^{-1} = \begin{bmatrix} 0.0054 & 0.0509 & -0.0066 \\ 0.0104 & -0.0186 & 0.0095 \\ -0.0193 & 0.0116 & 0.0344 \end{bmatrix}$$

53. $\begin{bmatrix} 44 & 21 & 18 & 6 \\ -2 & 10 & 15 & 5 \\ 21 & 12 & -12 & 4 \\ -8 & -16 & 4 & 9 \end{bmatrix}$

First, enter the matrix A into [A] on your calculator. Set the calculator to round the entries to four places after the decimal.

Find the inverse matrix of [A].

```
[A]-1
[[.0249   -.0360...
 [-.0171 .0521  ...
 [.0206   .0081 ...
 [-.0175 .0570  ...
```

```
[A]-1
...60  -.0057 .005...
...1   .0292   -.03...
...1   -.0421 4.87...
...0   .0657   .061...
```

```
[A]-1
...057 .0059      ]
...92   -.0305    ]
...421 4.8764E-4]
...57   .0619     ]]
```

Note that `4.8764E-4` is a number in scientific notation on the calculator. The number `4.8764E-4` is equivalent to 4.8764×10^{-4}.

Thus $A^{-1}=\begin{bmatrix} 0.0249 & -0.0360 & -0.0057 & 0.0059 \\ -0.0171 & 0.0521 & 0.0292 & -0.0305 \\ 0.0206 & 0.0081 & -0.0421 & 0.0005 \\ -0.0175 & 0.0570 & 0.0657 & 0.0619 \end{bmatrix}$.

55. $\begin{bmatrix} 3 & 0 & 2 & -1 & 3 \\ -2 & 1 & 2 & 3 & 0 \\ 2 & 2 & 1 & 1 & -1 \\ 1 & 2 & 0 & 2 & -3 \\ 4 & 0 & -1 & 1 & -1 \end{bmatrix}$

First, enter the matrix A into `[A]` on your calculator. Set the calculator to `Float`.

Find the inverse matrix of `[A]`, represent any decimal numbers in fraction form.

```
[A]-1▸Frac
[[1/4   -1/16 -9...
 [-3/2 1/8    49...
 [7/4   -3/16 -9...
 [-1/2 3/8    11...
 [-5/4 5/16   77...
```

```
[A]-1▸Frac
...  -9/32   5/16  ...
...  49/16   -21/8 ...
...  -91/32 47/16  ...
...  11/16   -7/8  ...
...  77/32   -41/16...
```

```
[A]-1▸Frac
...5/16     3/32  ]
...-21/8    5/16  ]
...47/16    -23/32]
...-7/8     7/16  ]
...-41/16 17/32 ]]
```

Thus, $A^{-1} = \begin{bmatrix} \frac{1}{4} & -\frac{1}{16} & -\frac{9}{32} & \frac{5}{16} & \frac{3}{32} \\ -\frac{3}{2} & \frac{1}{8} & \frac{49}{16} & -\frac{21}{8} & \frac{5}{16} \\ \frac{7}{4} & -\frac{3}{16} & -\frac{91}{32} & \frac{47}{16} & -\frac{23}{32} \\ -\frac{1}{2} & \frac{3}{8} & \frac{11}{16} & -\frac{7}{8} & \frac{7}{16} \\ -\frac{5}{4} & \frac{5}{16} & \frac{77}{32} & -\frac{41}{16} & \frac{17}{32} \end{bmatrix}$.

In Problems 57–60, use the idea behind Example 7 and either a graphing utility or EXCEL to solve each system of equations.

57. $\begin{cases} 25x + 61y - 12z = 10 \\ 18x - 12y + 7z = -9 \\ 3x + 4y - z = 12 \end{cases}$

Enter the coefficient matrix into [A] and the constants into the column matrix [B].

Remember that $X = A^{-1}B$, so find the product $A^{-1}B$.

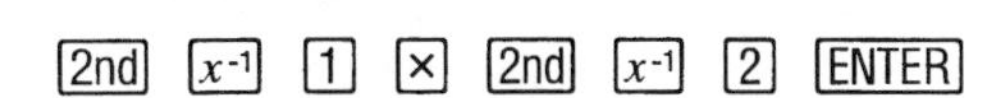

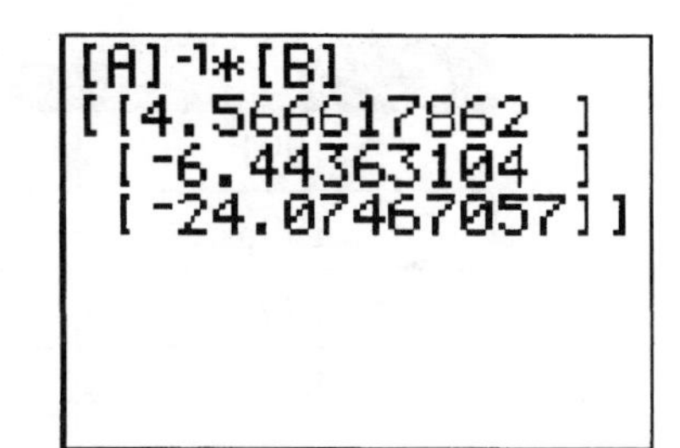

The solution to the system is $x = 4.5666$, $y = -6.4436$, and $z = -24.0747$.

59. $\begin{cases} 25x + 61y - 12z = 21 \\ 18x - 12y + 7z = 7 \\ 3x + 4y - z = -2 \end{cases}$

Enter the coefficient matrix into [A] and the constants into the column matrix [B].

Find the product $A^{-1}B$.

```
[A]-1*[B]
[[-1.187408492]
 [2.456808199 ]
 [8.265007321 ]]
```

The solution to the system is $x = -1.1874$, $y = 2.4568$, $z = 8.2650$.

Chapter 2 - Review

65. The U.S. Bureau of the Census publishes estimates of emigration from the United States to other countries. The following table lists the estimated emigration rates for selected years from 1990–2001. Use the data in the following questions.

Year	Emigrates
1991	252,000
1993	258,000
1996	267,000
1998	278,000
2000	287,000
2001	293,000

(f) Compute the line of best fit using a graphing utility and compare it to the line obtained in part (b).

We will use the procedure discussed in Exercise 7 of Section 1.4. Before entering the data for this problem, be sure to clear the lists L1 and L2. Enter the values for x into L1 and the values for y into L2. Remember to use 1 for 1991, 3 for 1993, etc.

L1	L2	L3 2
1	252000	------
3	258000	
6	267000	
8	278000	
10	287000	
11	293000	

L2(7) =

Find the line of best fit and store the equation in y1.

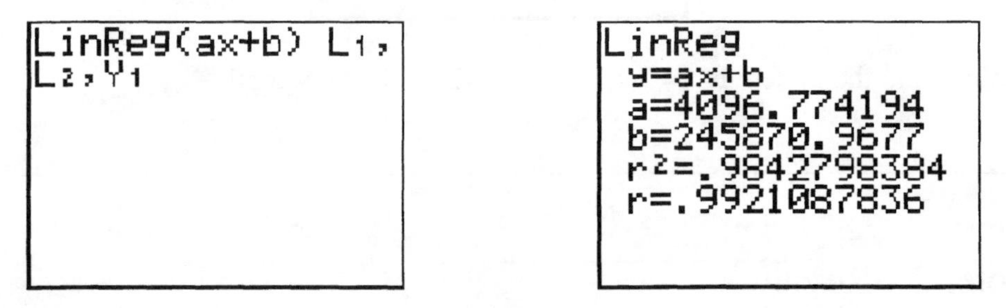

The line of best fit is $y = 4096.7742x + 245870.9677$.

Summary

The commands introduced in this chapter are:

ref(*matrixname*)

rref(*matrixname*)

matrixname^{-1}

▷Frac *expression*

Chapter 3 – Linear Programming: Geometric Approach

No Technology Problems in Chapter 3.

Chapter 4 – Linear Programming: Simplex Method

Chapter 4 Review

In problems 15–22, use each tableau to:

(a) Choose the pivot element and perform a pivot.

(b) Write the system of equations resulting from pivoting in part (a).

(c) Determine if the new tableau: (1) is the final tableau and, if so, write the solution; (2) requires additional pivoting and, if so, choose the new pivot element, or (3) indicates no solution exists for the problem.

21.

BV	P	x_1	x_2	x_3	s_1	s_2	s_3	RHS
s_1	0	−1	0	1	1	−1	0	7
x_2	0	−1	1	5	0	1	0	5
s_3	0	1	0	3	0	−5	1	3
P	1	−3	0	4	0	0	0	5

(a) Choose the pivot element and perform a pivot.

The pivot element is in row 3 and column 2.

The elementary row operations are built into your calculator. If you want to multiply a row of a matrix by a nonzero constant, the format for this operation is

`*row(`*constant*`,` *matrixname*`,` *rownumber*`)`

If you want to add a multiple of one row to another row of a matrix, the format for this operation is

`*row+(`*constant*`,` *matrixname*`,` *rowmultipled*`,` *rowaddedto*`)`

If you want to interchange two rows of a matrix, the format of the command is

`rowSwap(`*matrixname*, *rownumber*$_1$, *rownumber*$_2$`)`

If you want to add one row to another row of a matrix, the format for this operation is

`row+(`*matrixname*, *rownumber*, *rowaddedto*`)`

Enter tableau into the matrix `[A]` on your calculator.

Recall that we must first make the pivot a 1. In this case, the pivot is already 1. Next, we need to make the other entries in the pivot column 0. First, we can add row 3 to row 1, that is $R_1 = r_3 + r_1$. After each operation we must store the result in `[A]` to update the matrix for the next operation.

[2nd] [x⁻¹] [▶] [▲] [▲] [▲] [ENTER] [2nd] [x⁻¹] [1] [,] [3] [,] [1]

[)] [STO▶] [2nd] [x⁻¹] [1] [ENTER]

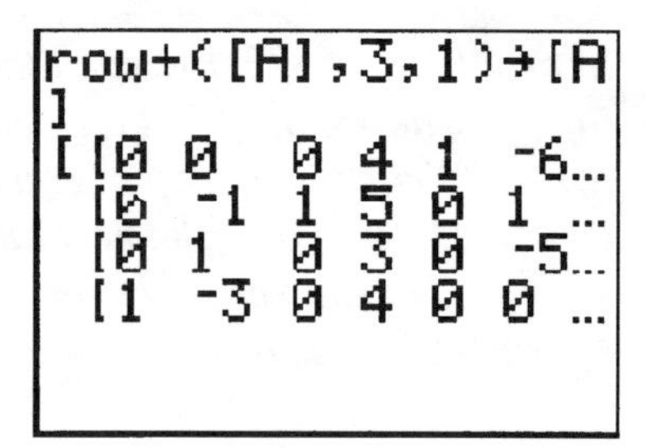

Scroll through the rest of the matrix.

```
row+([A],3,1)→[A
]
...0 4 1 -6 1 10]
...1 5 0 1  0 5 ]
...0 3 0 -5 1 3 ]
...0 4 0 0  0 5 ]]
```

Next, add row 3 to row 2, that is $R_2 = r_3 + r_2$.

```
row+([A],3,2)→[A
]
[[0 0  0 4 1 -6...
 [0 0  1 8 0 -4...
 [0 1  0 3 0 -5...
 [1 -3 0 4 0 0 ...
```

```
row+([A],3,2)→[A
]
...0 4 1 -6 1 10]
...1 8 0 -4 1 8 ]
...0 3 0 -5 1 3 ]
...0 4 0 0  0 5 ]]
```

Finally, we must add 3 times row 3 to row 4, that is $R_4 = 3r_3 + r_4$.

[2nd] [x^{-1}] [▶] [▲] [ENTER] [3] [,] [2nd] [x^{-1}] [1] [,] [3] [,] [4]

[)] [STO▶] [x^{-1}] [MATH] [1] [ENTER]

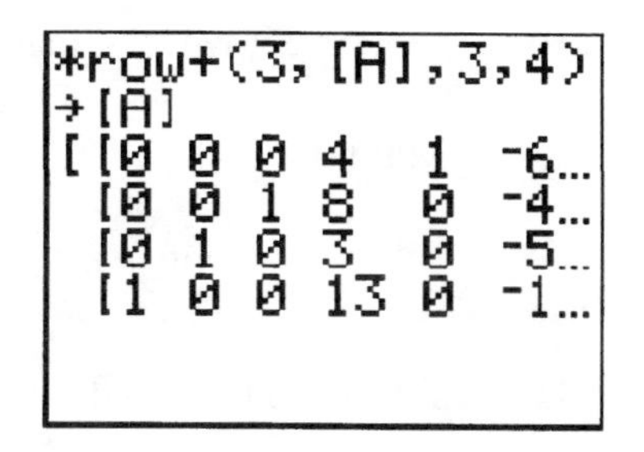

Scroll through the rest of the matrix.

```
*row+(3, [A],3,4)
→[A]
...4  1  -6   1  10]
...8  0  -4   1  8 ]
...3  0  -5   1  3 ]
...13 0  -15  3  14]]
```

The new simplex tableau is

BV	P	x_1	x_2	x_3	s_1	s_2	s_3	RHS
s_1	0	0	0	4	1	−6	1	10
x_2	0	0	1	8	0	−4	1	8
x_1	0	1	0	3	0	−5	1	3
P	1	0	0	13	0	−15	3	14

(b) Write the system of equations resulting from pivoting in part (a).

$$s_1 = 10 - 4x_3 + 6s_2 - s_3$$
$$x_2 = 8 - 8x_3 + 4s_2 - s_3$$
$$x_1 = 3 - 3x_3 + 5s_2 - s_3$$
$$P = 14 - 13x_3 + 15s_2 - 3s_3$$

(c) Determine if the new tableau: (1) is the final tableau and, if so, write the solution; (2) requires additional pivoting and, if so, choose the new pivot, or (3) indicates no solution exists for the problem.

The new tableau indicates there is no solution to the problem since all of the entries in column 6 are negative.

39. Solve Problem 35 using the duality principle.

From Problem 35, we find that the dual problem is

Maximize

$$P = 100y_1 + 50y_2$$

subject to the constraints

$$\begin{aligned} y_1 + 2y_2 &\le 5 \\ y_1 + \ \ y_2 &\le 4 \\ y_1 \qquad &\le 3 \\ y_1 \ge 0 \quad y_2 &\ge 0 \end{aligned}$$

The initial simplex tableau is

$$\begin{array}{c} \text{BV} \\ s_1 \\ s_2 \\ s_3 \\ P \end{array} \begin{array}{c} \begin{array}{ccccccc|c} P & y_1 & y_2 & s_1 & s_2 & s_3 & & \text{RHS} \end{array} \\ \left[\begin{array}{cccccc|c} 0 & 1 & 2 & 1 & 0 & 0 & 5 \\ 0 & 1 & 1 & 0 & 1 & 0 & 4 \\ 0 & 1 & 0 & 0 & 0 & 1 & 3 \\ \hline 1 & -100 & -50 & 0 & 0 & 0 & 0 \end{array}\right] \end{array}$$

Enter tableau into the matrix [A] on your calculator.

Find the first pivot. In this problem, the pivot is the third row, second column. Since the pivot is already a 1, we need to make the other entries in the second column 0. Use the following row operations:

$$\begin{aligned} R_1 &= -1r_3 + r_1 \\ R_2 &= -1r_3 + r_2 \\ R_4 &= 100r_3 + r_4 \end{aligned}$$

```
*row+(-1,[A],3,1
)→[A]
[[0 0     2   1 ...
 [0 1     1   0 ...
 [0 1     0   0 ...
 [1 -100  -50 0 ...
```

```
*row+(-1,[A],3,1
)→[A]
... 2    1 0 -1 2]
... 1    0 1 0  4]
... 0    0 0 1  3]
... -50 0 0 0   0]]
```

```
*row+(-1,[A],3,2
)→[A]
[[0 0     2   1 ...
 [0 0     1   0 ...
 [0 1     0   0 ...
 [1 -100  -50 0 ...
```

```
*row+(-1,[A],3,2
)→[A]
... 2    1 0 -1 2]
... 1    0 1 -1 1]
... 0    0 0 1  3]
... -50 0 0 0   0]]
```

```
*row+(100,[A],3,
4)→[A]
[[0 0 2   1 0 -...
 [0 0 1   0 1 -...
 [0 1 0   0 0 1...
 [1 0 -50 0 0 1...
```

```
*row+(100,[A],3,
4)→[A]
...  1 0 -1  2   ]
...  0 1 -1  1   ]
...  0 0 1   3   ]
...0 0 0 100 300]]
```

The new simplex tableau is

BV	P	y_1	y_2	s_1	s_2	s_3	RHS
s_1	0	0	2	1	0	−1	2
s_2	0	0	1	0	1	−1	1
y_1	0	1	0	0	0	1	3
P	1	0	−50	0	0	100	300

Since there is still a negative entry in the objective row, we must pivot again. This time, the pivot is in column three, but could be in row one or in row two. We can use either row, so let's use row two. Since the pivot is already a 1, we need to make the other entries in the second column 0. Use the following row operations:

$$R_1 = -2r_2 + r_1$$
$$R_4 = 50r_2 + r_4$$

```
*row+(-2,[A],2,1
)→[A]
[[0 0 0   1 -2 ...
 [0 0 1   0 1  ...
 [0 1 0   0 0  ...
 [1 0 -50 0 0  ...
```

```
*row+(-2,[A],2,1
)→[A]
... 1 -2 1   0  ]
... 0 1  -1  1  ]
... 0 0  1   3  ]
... 0 0  100 300]]
```

```
*row+(50,[A],2,4
)→[A]
[[0 0 0 1 -2 1 ...
 [0 0 1 0 1  -1...
 [0 1 0 0 0  1 ...
 [1 0 0 0 50 50...
```

```
*row+(50,[A],2,4
)→[A]
...0 1 -2 1  0  ]
...1 0 1  -1 1  ]
...0 0 0  1  3  ]
...0 0 50 50 350]]
```

The final simplex tableau is

BV	P	y_1	y_2	s_1	s_2	s_3	RHS
s_1	0	0	0	1	−2	1	0
y_2	0	0	1	0	1	−1	1
y_1	0	1	0	0	0	1	3
P	1	0	0	0	50	50	350

The optimal solution is $P = 350$, $y_1 = 3$, and $y_2 = 1$. Using the duality principle, we find that $C = 350$ when $x_1 = 0$, $x_2 = 50$, and $x_3 = 50$.

In Problems 41–46, use the mixed constraints method to solve each linear programming problem.

41. Maximize

$$P = 3x_1 + 5x_2$$

subject to the constraints

$$x_1 + x_2 \geq 2$$
$$2x_1 + 3x_2 \leq 12$$
$$3x_1 + 2x_2 \leq 12$$
$$x_1 \geq 0 \quad x_2 \geq 0$$

The initial simplex tableau is

BV	P	x_1	x_2	s_1	s_2	s_3	RHS
s_1	0	−1	−1	1	0	0	−2
s_2	0	2	3	0	1	0	12
s_3	0	3	2	0	0	1	12
P	1	−3	−5	0	0	0	0

Enter tableau into the matrix `[A]` on your calculator.

We must first eliminate the negative in the RHS. The pivot will be in row one since the negative entry in RHS is in row one. The pivot column will be in column two, since it is the first negative entry we come across as we read from left to right. Use the following row operations:

$$R_1 = -1r_1$$
$$R_2 = -2r_1 + r_2$$
$$R_3 = -3r_1 + r_3$$
$$R_4 = 3r_1 + r_4$$

```
*row(-1,[A],1)→[
A]
[[0 1   1  -1 0 ...
 [0 2   3  0  1 ...
 [0 3   2  0  0 ...
 [1 -3 -5  0  0 ...
```

```
*row(-1,[A],1)→[
A]
... 1   -1 0 0 2 ]
... 3   0  1 0 12]
... 2   0  0 1 12]
... -5  0  0 0 0 ]]
```

```
*row+(-2,[A],1,2
)→[A]
[[0 1  1  -1 0 ...
 [0 0  1  2  1 ...
 [0 3  2  0  0 ...
 [1 -3 -5 0  0 ...
```

```
*row+(-2,[A],1,2
)→[A]
... 1  -1 0 0 2 ]
... 1  2  1 0 8 ]
... 2  0  0 1 12]
... -5 0  0 0 0 ]]
```

```
*row+(-3,[A],1,3
)→[A]
[[0 1  1  -1 0 ...
 [0 0  1  2  1 ...
 [0 0  -1 3  0 ...
 [1 -3 -5 0  0 ...
```

```
*row+(-3,[A],1,3
)→[A]
...  1  -1 0 0 2]
...  1  2  1 0 8]
...  -1 3  0 1 6]
...3 -5 0  0 0 0]]
```

```
*row+(3,[A],1,4)
→[A]
[[0 1 1  -1 0 0...
 [0 0 1  2  1 0...
 [0 0 -1 3  0 1...
 [1 0 -2 -3 0 0...
```

```
*row+(3,[A],1,4)
→[A]
...1 1  -1 0 0 2]
...0 1  2  1 0 8]
...0 -1 3  0 1 6]
...0 -2 -3 0 0 6]]
```

The new simplex tableau is

BV	P	x_1	x_2	s_1	s_2	s_3	RHS
x_1	0	1	1	−1	0	0	2
s_2	0	0	1	2	1	0	8
s_3	0	0	−1	3	0	1	6
P	1	0	−2	−3	0	0	6

Since there are still negative entries in the objective row, we must pivot again. This time, the pivot is in row three, column four. Use the following row operations:

$$R_3 = \tfrac{1}{3} r_3$$
$$R_1 = r_3 + r_1$$
$$R_2 = -2r_3 + r_2$$
$$R_4 = 3r_3 + r_4$$

```
*row(1/3,[A],3)→
[A]
[[0 1 1           ...
 [0 0 1           ...
 [0 0 -.3333333...
 [1 0 -2          ...
```

Since some of the entries are decimal, use ▷Frac to convert the decimals to fractions.

```
Ans▸Frac
[[0 1 1      -1 0...
 [0 0 1      2  1...
 [0 0 -1/3 1  0...
 [1 0 -2     -3 0...
```

```
Ans▸Frac
...    -1 0 0    2]
...    2  1 0    8]
.../3 1  0 1/3 2]
...    -3 0 0    6]]
```

```
row+([A],3,1)→[A
]
[[0 1 .66666666...
 [0 0 1           ...
 [0 0 -.3333333...
 [1 0 -2          ...
```

```
Ans▸Frac
[[0 1 2/3   0  0...
 [0 0 1     2  1...
 [0 0 -1/3 1  0...
 [1 0 -2    -3 0...
```

```
Ans▸Frac
...3  0  0 1/3 4]
...    2  1 0    8]
.../3 1  0 1/3 2]
...    -3 0 0    6]]
```

```
*row+(-2,[A],3,2
)→[A]
[[0 1 .66666666...
 [0 0 1.6666666...
 [0 0 -.3333333...
 [1 0 -2          ...
```

```
Ans▸Frac
[[0 1 2/3  0  0...
 [0 0 5/3  0  1...
 [0 0 -1/3 1  0...
 [1 0 -2   -3 0...
```

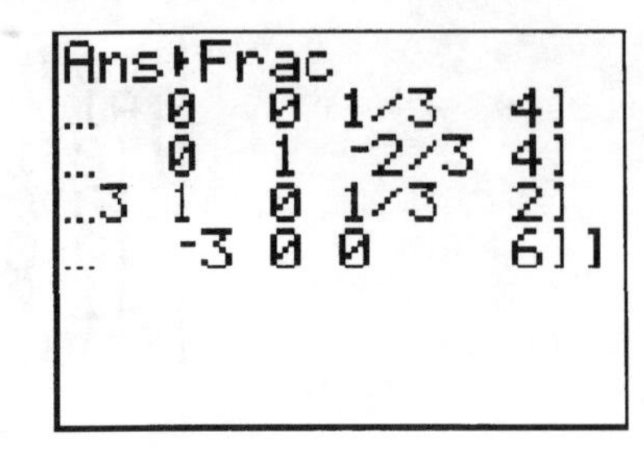


```
*row+(3,[A],3,4)
→[A]
[[0 1 .66666666...
 [0 0 1.6666666...
 [0 0 -.3333333...
 [1 0 -3        ...
```

```
Ans▸Frac
[[0 1 2/3  0 0 ...
 [0 0 5/3  0 1 ...
 [0 0 -1/3 1 0 ...
 [1 0 -3   0 0 ...
```

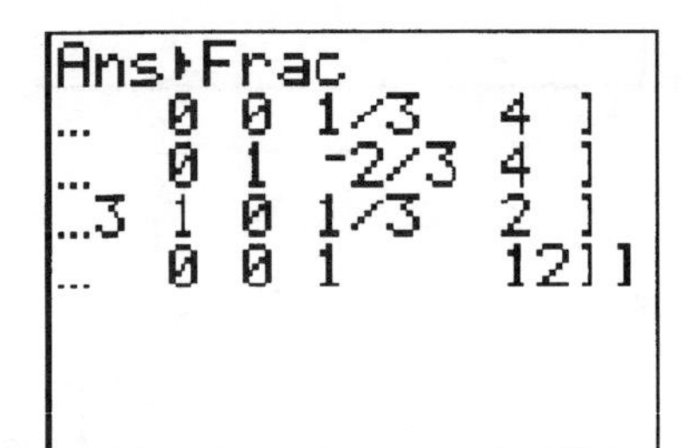

The new simplex tableau is

BV	P	x_1	x_2	s_1	s_2	s_3	RHS
x_1	0	1	$\frac{2}{3}$	0	0	$\frac{1}{3}$	4
s_2	0	0	$\frac{5}{3}$	0	1	$-\frac{2}{3}$	4
s_1	0	0	$-\frac{1}{3}$	1	0	$\frac{1}{3}$	2
P	1	0	-3	0	0	1	12

Since there is still a negative entry in the objective row, we must pivot again. This time, the pivot is in row two, column three. Use the following row operations:

$$R_2 = \tfrac{3}{5}r_2$$
$$R_1 = -\tfrac{2}{3}r_2 + r_1$$
$$R_3 = \tfrac{1}{3}r_2 + r_3$$
$$R_4 = 3r_2 + r_4$$

```
*row(3/5,[A],2)→
[A]
[[0 1 .66666666...
 [0 0 1         ...
 [0 0 -.3333333...
 [1 0 -3        ...
```

```
*row+(-2/3,[A],2
,1)→[A]
[[0 1 0         ...
 [0 0 1         ...
 [0 0 -.3333333...
 [1 0 -3        ...
```

```
*row+(1/3,[A],2,
3)→[A]
[[0 1 0  0 -.4 ...
 [0 0 1  0 .6  ...
 [0 0 0  1 .2  ...
 [1 0 -3 0 0   ...
```

```
*row+(3,[A],2,4)
→[A]
[[0 1 0 0 -.4 ....
 [0 0 1 0 .6  -...
 [0 0 0 1 .2  ....
 [1 0 0 0 1.8 -...
```

```
Ans▸Frac
[[0 1 0 0 -2/5 ...
 [0 0 1 0 3/5  ...
 [0 0 0 1 1/5  ...
 [1 0 0 0 9/5  ...
```

```
Ans▸Frac
...2/5 3/5  12/5]
.../5  -2/5 12/5]
.../5  1/5  14/5]
.../5  -1/5 96/5]]
```

The new simplex tableau is

$$\begin{array}{c|cccccc|c|}
\text{BV} & P & x_1 & x_2 & s_1 & s_2 & s_3 & \text{RHS} \\
\hline
x_1 & 0 & 1 & 0 & 0 & -\frac{2}{5} & \frac{3}{5} & \frac{12}{5} \\
x_2 & 0 & 0 & 1 & 0 & \frac{3}{5} & -\frac{2}{5} & \frac{12}{5} \\
s_1 & 0 & 0 & 0 & 1 & \frac{1}{5} & \frac{1}{5} & \frac{14}{5} \\
\hline
P & 1 & 0 & 0 & 0 & \frac{9}{5} & -\frac{1}{5} & \frac{96}{5}
\end{array}$$

Since there is still a negative entry in the objective row, we must pivot again. This time, the pivot is in row one, column six. Use the following row operations:

$$R_1 = \tfrac{5}{3} r_1$$
$$R_2 = \tfrac{2}{5} r_1 + r_2$$
$$R_3 = -\tfrac{1}{5} r_1 + r_3$$
$$R_4 = \tfrac{1}{5} r_1 + r_4$$

```
*row(5/3,[A],1)→
[A]
[[0 1.666666667...
 [0 0          ...
 [0 0          ...
 [1 0          ...
```

```
*row+(2/5,[A],1,
2)→[A]
[[0 1.666666667...
 [0 .6666666667...
 [0 0          ...
 [1 0          ...
```

```
*row+(-1/5,[A],1
,3)→[A]
[[0 1.666666667...
 [0 .6666666667...
 [0 -.333333333...
 [1 0          ...
```

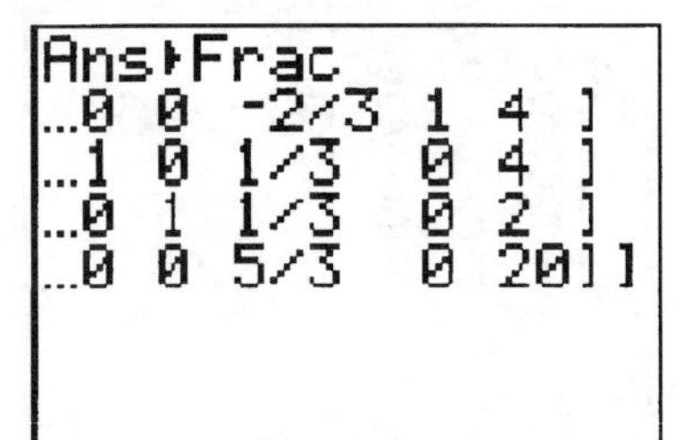


```
Ans▸Frac
[[0 5/3  0 0 -2...
 [0 2/3  1 0 1/...
 [0 -1/3 0 1 1/...
 [1 1/3  0 0 5/...
```

```
Ans▸Frac
...0 0 -2/3 1 4 ]
...1 0 1/3  0 4 ]
...0 1 1/3  0 2 ]
...0 0 5/3  0 20]]
```

The final simplex tableau is

BV	P	x_1	x_2	s_1	s_2	s_3	RHS
s_3	0	$\frac{5}{3}$	0	0	$-\frac{2}{3}$	1	4
x_2	0	$\frac{2}{3}$	1	0	$\frac{1}{3}$	0	4
s_1	0	$-\frac{1}{3}$	0	1	$\frac{1}{3}$	0	2
P	1	$\frac{1}{3}$	0	0	$\frac{5}{3}$	0	20

The maximum value of P is 20, and it is achieved when $x_1 = 0$, $x_2 = 4$, $s_1 = 2$, $s_2 = 0$, and $s_3 = 4$

43. Minimize

$$C = 2x_1 + 3x_2$$

subject to the constraints

$$x_1 + x_2 \geq 3$$
$$x_1 + x_2 \leq 9$$
$$x_1 \geq 0 \quad x_2 \geq 0$$

Rewriting our objective function as $P = -C = -2x_1 - 3x_2$, writing each constraint with $\leq$, and introducing nonnegative slack variables to form equations, we can obtain the following system:

$$\begin{aligned} -x_1 - x_2 + s_1 \quad &= -3 \\ x_1 + x_2 \quad + s_2 &= \ \ 9 \\ x_1 \geq 0 \quad x_2 \geq 0 \quad s_1 \geq 0 \quad s_2 &\geq 0 \end{aligned}$$

The objective function is: $P = -2x_1 - 3x_2$

The initial simplex tableau is

BV	P	x_1	x_2	s_1	s_2	RHS
s_1	0	−1	−1	1	0	−3
s_2	0	1	1	0	1	9
P	1	2	3	0	0	0

Enter tableau into the matrix [A] on your calculator.

We must first eliminate the negative in the RHS. The pivot will be in row one since the negative entry in RHS is in row one. The pivot column will be in column two, since it is the first negative entry we come across as we read from left to right. Use the following row operations:

$$R_1 = -1r_1$$
$$R_2 = -1r_1 + r_2$$
$$R_3 = -2r_1 + r_3$$

```
*row(-1,[A],1)→[
A]
[[0 1 1 -1 0 3]
 [0 1 1 0  1 9]
 [1 2 3 0  0 0]]
```

```
*row+(-1,[A],1,2
)→[A]
[[0 1 1 -1 0 3]
 [0 0 0 1  1 6]
 [1 2 3 0  0 0]]
```

```
*row+(-2,[A],1,3
)→[A]
[[0 1 1 -1 0 3 ...
 [0 0 0 1  1 6 ...
 [1 0 1 2  0 -6...
```

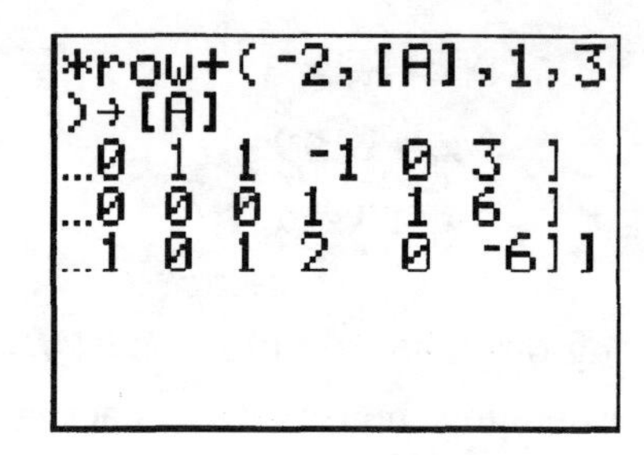

The new simplex tableau is

BV	P	x_1	x_2	s_1	s_2	RHS
x_1	0	1	1	-1	0	3
s_2	0	0	0	1	1	6
P	1	0	1	2	0	-6

This is the final tableau. The maximum value of P is -6, so the minimum value of C is 6. This occurs when $x_1 = 3$, $x_2 = 0$, $s_1 = 0$, and $s_2 = 6$.

Summary

The commands introduced in this chapter are:

`*row(`*constant*, *matrixname*, *rownumber*`)`

`*row+(`*constant*, *matrixname*, *rowmultipled*, *rowadded*`)`

`rowSwap(`*matrixname*, *rownumber*$_1$, *rownumber*$_2$`)`

`row+(`*matrixname*, *rownumber*, *rowaddedto*`)`

Chapter 5 – Finance

Section 5.5 Annuities and Amortization Using Recursive Sequences

1. **Credit Card Debt** John has a balance of \$3000 on his credit card that charges 1% interest per month on any unpaid balance. John can afford to pay \$100 toward the balance each month. His balance each month after making a \$100 payment is given by the recursively defined sequence

$$B_0 = \$3000\,,\ B_n = 1.01B_{n-1} - 100$$

(b) Using a graphing utility, determine when John's balance will be below \$2000. How many of the \$100 payments have been made?

Before we can enter the information for the sequence, we must first put the calculator in sequence mode.

[MODE] [▼] [▼] [▼] [▶] [▶] [▶] [ENTER]

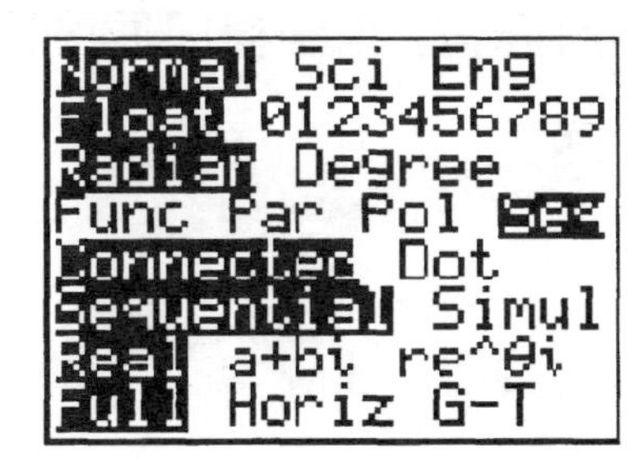

Enter the recursive definition of the sequence in the function editor. The calculator uses *n* for the variable and the letters u, v, and w for the names of the sequences. Note that the variable *n* is found on the [X,T,Θ,n] key, and the letters u, v, and w are above the [7], [8], and [9] keys, respectively.

[Y=] [▲] [0] [ENTER] [1] [.] [0] [1] [2nd] [7] [(] [X,T,Θ,n] [−] [1] [)]

[−] [1] [0] [0] [ENTER] [3] [0] [0] [0] [ENTER]

Plot1 Plot2 Plot3
nMin=0
u(n)=1.01u(n−1)−100
u(nMin)={3000}
v(n)=
v(nMin)=
w(n)=

Next we must set up the table feature so we can choose the values for *n*.

[2nd] [WINDOW] [▼] [▼] [▶] [ENTER]

Now, create the table.

[2nd] [GRAPH] [0] [ENTER]

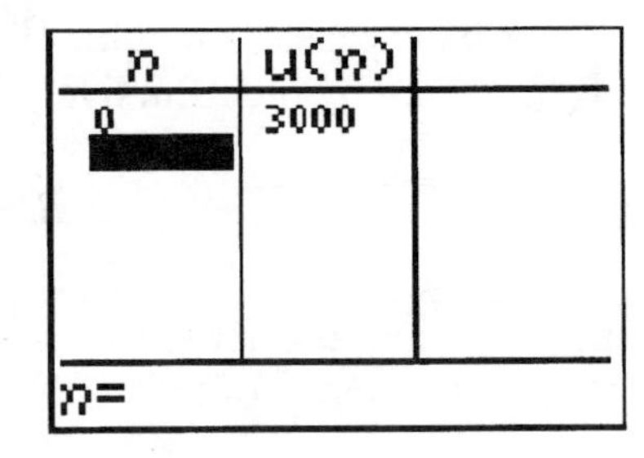

[1] [ENTER] [2] [ENTER] [3] [ENTER] [4] [ENTER] [5] [ENTER] [6] [ENTER]

n	u(n)
0	3000
1	2930
2	2859.3
3	2787.9
4	2715.8
5	2642.9
6	2569.4

n=6

Continue to enter values. Be sure to move back the top of the first column.

[▲] [▲] [▲] [▲] [▲] [▲] [7] [ENTER] [8] [ENTER] [9] [ENTER] [1] [0] [ENTER] [1] [1] [ENTER] [1] [2] [ENTER] [1] [3] [ENTER]

n	u(n)
7	2495.1
8	2420
9	2344.2
10	2267.6
11	2190.3
12	2112.2
13	2033.3

n=13

Continue to enter values.

n	$u(n)$
14	1953.7
15	1873.2
16	1791.9
17	1709.9
18	1627
19	1543.2
20	1458.7

$n=20$

After the 14th payment, the balance is below \$2000.

(c) Using a graphing utility, determine when John will pay off the balance. What is the total of all of the payments.

Continue to enter values in the table. You can use the DEL key to delete unwanted entries from the table.

n	$u(n)$
21	1373.3
22	1287
23	1199.9
24	1111.9
25	1023
26	933.21
27	842.54

$n=27$

n	$u(n)$
28	750.96
29	658.47
30	565.06
31	470.71
32	375.42
33	279.17
34	181.96

$n=34$

n	$u(n)$
35	83.781
36	⁻15.38

$n=$

John will have to make 36 payments.

John's final payment should be \$84.62 (since he overpaid by \$15.38), so John's total of all payments is $\$100 \times 35 + \$84.62 = \$3,584.62$.

Be sure to set your calculator back in Func (function) mode before you try to graph any functions. To set your calculator in Func mode, press [MODE], move the cursor down to Func and press [ENTER].

3. **Trout Population** A pond currently has 2000 trout in it. A fish hatchery decides to add an additional 20 trout each month. In addition, it is known that the trout population is growing 3% per month. The size of the population after n months is given by the recursively defined sequence

$$p_0 = 2000,\ p_n = 1.03p_{n-1} + 20$$

(b) Using a graphing utility, determine how long it will be before the trout population reaches 5000.

Enter the recursive definition of the sequence in the function editor.

```
Plot1 Plot2 Plot3
nMin=0
\u(n)=1.03u(n-1)
+20
 u(nMin)={2000}
\v(n)=
 v(nMin)=
\w(n)=
```

Using the table feature, enter values for n until the value of the sequence reaches 5000.

n	u(n)
0	2000
1	2080
2	2162.4
3	2247.3
4	2334.7
5	2424.7
6	2517.5

n=6

n	u(n)
7	2613
8	2711.4
9	2812.7
10	2917.1
11	3024.6
12	3135.4
13	3249.4

n=13

n	u(n)
14	3366.9
15	3487.9
16	3612.6
17	3740.9
18	3873.2
19	4009.3
20	4149.6

n=20

n	u(n)
21	4294.1
22	4442.9
23	4596.2
24	4754.1
25	4916.7
26	5084.2

n=

The population will reach 5000 in the 26th month.

Be sure to set your calculator back in Func (function) mode before you try to graph any functions. To set your calculator in Func mode, press [MODE], move the cursor down to Func and press [ENTER].

5. **Roth IRA** On January 1, 1999, Bob decides to place \$500 at the end of each quarter into a Roth Individual Retirement Account.

(b) How long will it be before the value of the account exceeds \$100,000?

Enter the recursive definition of the sequence in the function editor.

```
Plot1 Plot2 Plot3
nMin=0
u(n)=1.02u(n-1)
+500
u(nMin)={0}
v(n)=
v(nMin)=
w(n)=
```

Using the table feature, enter values for n. until you pass 100000.

n	u(n)
77	89856
78	92153
79	94496
80	96886
81	99324
82	101810

n=

After the 82nd payment.

(c) What will be the value of the account in 25 years when Bob retires?

When Bob retires, he will have made 100 payments.

n	$u(n)$
100	156116

$u(n)=156116.153$

The value is $156,116.15.

7. **Home Loan** Bill and Laura borrowed $150,000 at 6% per annum compounded monthly for 30 years to purchase a home. Their monthly payment is determined to be $899.33.

(c) Using a graphing utility, create a table showing Bill and Laura's balance after each monthly payment.

Enter the recursive definition of the sequence in the function editor.

```
Plot1 Plot2 Plot3
nMin=0
\u(n)=(1+.06/12)
u(n-1)-899.33
 u(nMin)=(15000...
\v(n)=
 v(nMin)=
\w(n)=
```

Using the table feature, enter values for n.

n	$u(n)$
0	150000
1	149851
2	149701
3	149550
4	149398
5	149246
6	149093

$n=6$

n	$u(n)$
7	148939
8	148784
9	148629
10	148473
11	148316
12	148158
13	147999

$n=13$

n	$u(n)$
14	147840
15	147680
16	147519
17	147357
18	147195
19	147031
20	146867

$n=20$

To complete the table, enter values from 21 to 360 for n.

(d) Using a graphing utility, determine when Bill and Laura's balance will be below $140,000.

As you continue to enter values for n you will obtain the following.

n	$u(n)$
56	140377
57	140180
58	139981
59	139782
60	139581
61	139380
62	139177

$n=62$

After the 58th payment.

(e) Using a graphing utility, determine when Bill and Laura will pay off the balance.

As you continue to enter values for n you will obtain the following.

n	$u(n)$
357	2667.1
358	1781.1
359	890.65
360	-4.231

$n=$

After the 360th payment.

(g) Suppose that Bill and Laura decide to pay an additional $100 each month on their loan. Answer parts (a) to (f) under this scenario.

(c) Using a graphing utility, create a table showing Bill and Laura's balance after each monthly payment.

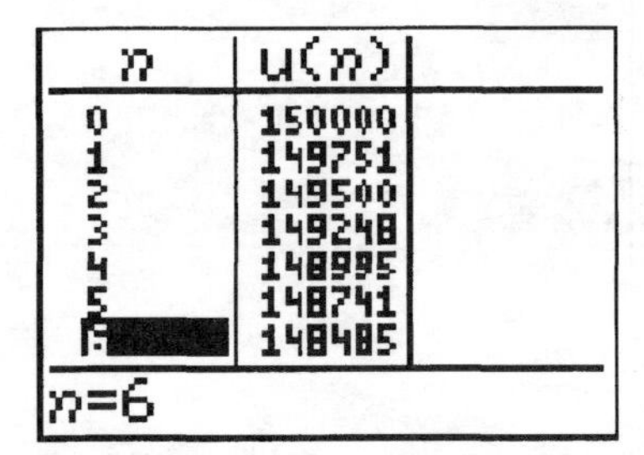

n	$u(n)$
0	150000
1	149751
2	149500
3	149248
4	148995
5	148741
6	148485

$n=6$

n	$u(n)$
7	148228
8	147970
9	147711
10	147450
11	147188
12	146924
13	146660

$n=13$

n	$u(n)$
14	146394
15	146126
16	145858
17	145588
18	145316
19	145043
20	144769

$n=20$

To complete the table, enter values from 21 on for n.

(d) Using a graphing utility, determine when Bill and Laura's balance will be below $140,000.

As you continue to enter values for n you will obtain the following.

n	$u(n)$
35	140489
36	140192
37	139894
38	139594
39	139293
40	138990
41	138685

$n=41$

After the 37th payment.

(e) Using a graphing utility, determine when Bill and Laura will pay off the balance.

As you continue to enter values for n you will obtain the following.

n	u(n)
273	5267.5
274	4294.6
275	3316.7
276	2333.9
277	1346.3
278	353.69
279	-643.9

n=279

After the 279th payment.

(h) Is it worthwhile for Bill and Laura to pay the additional \$100. Explain.

Yes. If Bill and Laura make a monthly payment of \$899.33, then the total interest paid is \$173,758.80. If Bill and Laura make a monthly payment of \$999.33, then the total interest paid is \$128,169.20. Bill and Laura saved \$45,589.60 in interest payments.

```
173758.80-128169
.20
                45589.6
```

Chapter 5 Review

In Problems 1 – 9, calculate the indicated quantity.

1. 3% of 500

[.] [0] [3] [×] [5] [0] [0] [ENTER]

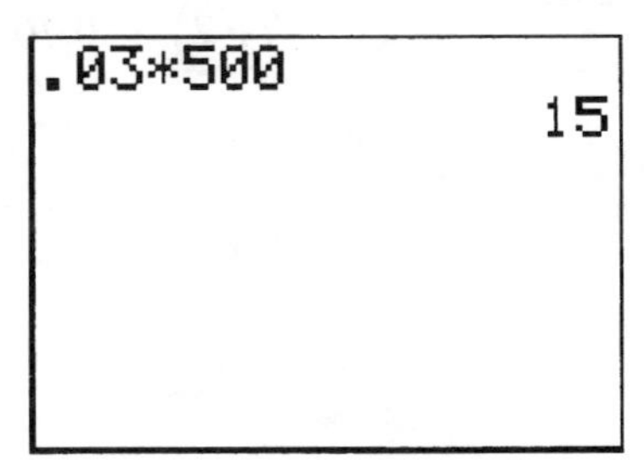

3% of 500 is 15.

5. What percent of 350 is 75?

[7] [5] [÷] [3] [5] [0] [ENTER]

```
75/350
          .2142857143
```

75 is approximately 21.43% of 350.

9. 11 is 0.5% of what number?

[1] [1] [÷] [.] [0] [0] [5] [ENTER]

```
11/.005
                 2200
```

11 is 0.5% of 2200.

15. **Loan Amount** Warren needs $15,000 for a new machine for his auto repair shop. He obtains a 2-year discounted loan at 12% interest. How much must he repay to settle his debt?

To find the amount Warren must repay, solve the equation $15000 = A\left(1-(.12)(2)\right)$ for A.

Thus, $A = \dfrac{15000}{1-(.12)(2)} = \dfrac{15000}{0.76}$.

15000/.76
19736.84211

[1] [5] [0] [0] [0] [÷] [.] [7] [6] [ENTER]

Warren must repay $19,736.84.

17. Find the amount of an investment of $100 after 2 years and 3 months at 10% compounded monthly.

Note that 2 years and 3 months is equivalent to $2+\frac{3}{12}=2.25$ years.

100(1+.1/12)^(12
*2.25)
125.115569

[1] [0] [0] [(] [1] [+] [.] [1] [÷] [1] [2] [)] [^] [(] [1] [2] [×] [2]

[.] [2] [5] [)] [ENTER]

The amount of the investment is $125.12.

23. **Doubling Money** What annual rate of interest will allow an investment to double in 12 years?

Solving the equation $2=(1+r)^{12}$ for r we obtain $r=\sqrt[12]{2}-1$.

To find a root, other than the square root, of a number we use the $\sqrt[x]{\ }$ function. This command is found in the [MATH] menu. To find the nth root of x the format is

$$n \sqrt[x]{\ } x$$

[1] [2] [MATH] [5] [(] [2] [)] [−] [1] [ENTER]

```
12ˣ√(2)-1
          .0594630944
```

Thus, $r \approx 0.0595$, or 5.95%.

27. **Saving for a House** Mr. and Mrs. Corey are newlyweds and want to purchase a home, but they need a down payment of $40,000. If they want to buy their home in 2 years, how much should they save each month in their savings account that pays 3% per annum compounded monthly?

The monthly is payment given by $PMT = \dfrac{40000}{\left(\dfrac{\left(1+\dfrac{.03}{12}\right)^{24}-1}{\dfrac{.03}{12}}\right)}$.

[4] [0] [0] [0] [0] [÷] [(] [(] [(] [1] [+] [.] [0] [3] [÷] [1] [2] [)]

[^] [2] [4] [−] [1] [)] [÷] [(] [.] [0] [3] [÷] [2] [)] [)] [ENTER]

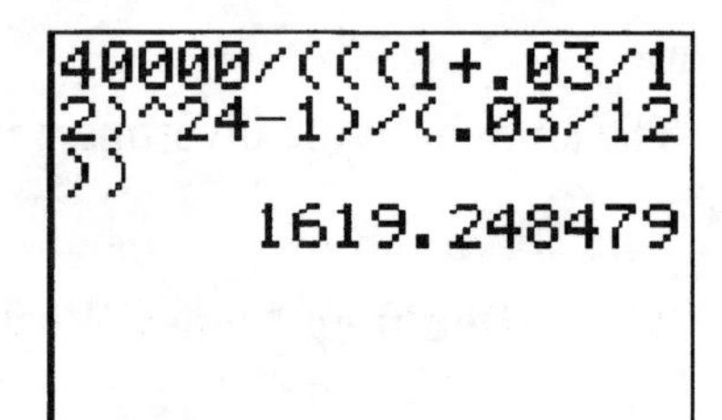

The monthly payment is $1,619.25.

29. **House Mortgage** Mr. and Mrs. Ostedt have just purchased an $400,000 home and made a 25% down payment. The balance can be amortized at 10% for 25 years.

    (a) What are the monthly payments?

The monthly is payment given by $PMT = 300000\frac{\left(\frac{.10}{12}\right)}{1-\left(1+\frac{.10}{12}\right)^{-300}}$.

[3] [0] [0] [0] [0] [0] [(] [(] [.] [1] [0] [÷] [1] [2] [)] [÷] [(] [1] [-] [(] [1] [+] [.] [1] [0] [÷] [1] [2] [)] [^] [(-)] [3] [0] [0] [)] [)] [ENTER]

```
300000((.10/12)/
(1-(1+.10/12)^-3
00))
       2726.102237
```

The monthly payment is \$2,726.10.

(b) How much interest will be paid?

[3] [0] [0] [×] [2] [7] [2] [6] [.] [1] [0] [-] [3] [0] [0] [0] [0] [0] [ENTER]

```
300*2726.10-3000
00
           517830
```

They will pay \$517,830 in interest.

(c) What is their equity after 5 years?

The amount still owed on the loan is given by $2726.10\frac{1-\left(1+\frac{.10}{12}\right)^{-240}}{\left(\frac{.10}{12}\right)}$.

[2] [7] [2] [6] [.] [1] [0] [(] [(] [1] [−] [(] [1] [+] [.] [1] [0] [÷]

[1] [2] [)] [^] [(-)] [2] [4] [0] [)] [÷] [(] [.] [1] [0] [÷] [1] [2] [)]

[)] [ENTER]

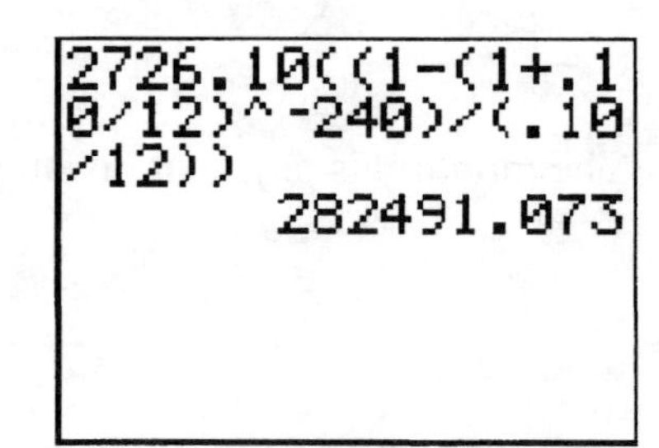

Thus, they still owe $282,491.07. The amount paid off is

```
300000-282491.07
          17508.93
```

Their total equity is the amount paid off plus the down payment.

```
100000+17508.93
         117508.93
```

Thus, their total equity is $117,508.93.

33. **Depletion Problem** How much should Mr. Graff pay for a gold mine expected to yield an annual return of $20,000 and to have a life expectancy of 20 years, if he wants to have a 15% annual return on his investment and he can set up a sinking fund that earns 10% a year?

Let x represent the amount that Mr. Graff should pay of the gold mine. We obtain the equations

$$x = \text{SFC}\frac{(1+.10)^{20}-1}{.10} \qquad \text{ROI} = .15x$$

Solve the first equation for SFC.

```
((1+.10)^20-1)/.
10
         57.27499949
1/57.27499949
        .0174596248
```

Thus, we obtain $\text{SFC} = 0.0174596248x$.

Since we know that $\text{SFC} + \text{ROI} = 20000$, we get the equation

$$.0.0174596248x + 0.15x = 200000$$

Solve this equation for x.

```
.0174596248+.15
        .1674596248
20000/.167459624
8
         119431.7736
```

Thus, we find that $x = 199{,}431.77$.

Mr. Graff should pay $199,431.77 for the gold mine.

45. **Trust Fund Payouts** John is the beneficiary of a trust fund set up for him by his grandparents. If the trust fund amounts to $20,000 earning 8% compounded semiannually and he is to receive the money in equal semiannual installments for the next 15 years, how much will he receive each 6 months?

The monthly is payment given by $PMT = 20000\dfrac{\left(\frac{.08}{2}\right)}{1-\left(1+\frac{.08}{2}\right)^{-30}}$.

```
20000((.08/2)/(1
-(1+.08/2)^-30))
      1156.601983
```

The monthly installment is \$1,156.60.

49. **Buying a Car** A student borrowed \$4000 from a credit union toward purchasing a car. The interest rate on such a loan is 14% compounded quarterly, with payments due every quarter. The student wants to pay off the loan in 4 years. Find the quarterly payment.

The monthly is payment given by $PMT = 4000\frac{\left(\frac{.14}{4}\right)}{1-\left(1+\frac{.14}{4}\right)^{-16}}$.

```
4000((.14/4)/(1-
(1+.14/4)^-16))
      330.7393226
```

The monthly installment is \$330.74.

Summary

The command introduced in this chapter is:

$$n \sqrt[x]{\ } x$$

Chapter 6 – Sets; Counting Techniques

Chapter 6 Review

In Problems 39 –58, evaluate the expression.

41. $\frac{7!}{4!}$

You can evaluate factorials using the ! command found in the PRB sub menu of the [MATH] menu.

[7] [MATH] [◄] [4] [÷] [4] [MATH] [◄] [4] [ENTER]

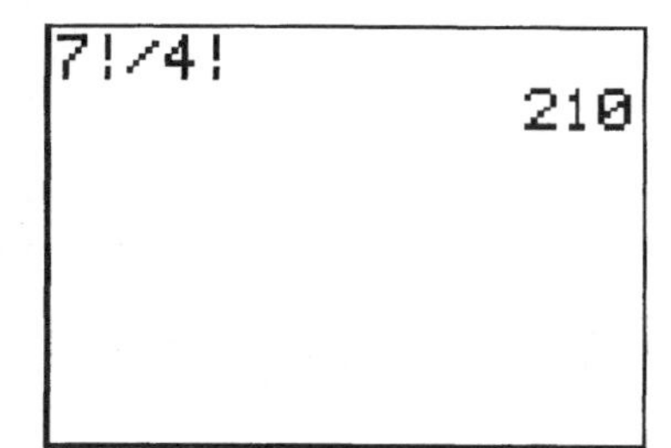

Thus, $\frac{7!}{4!} = 210$

51. $C(10,2)$

You can evaluate combinations using the nCr command found in the PRB sub menu of the [MATH] menu. The format is

n nCr r

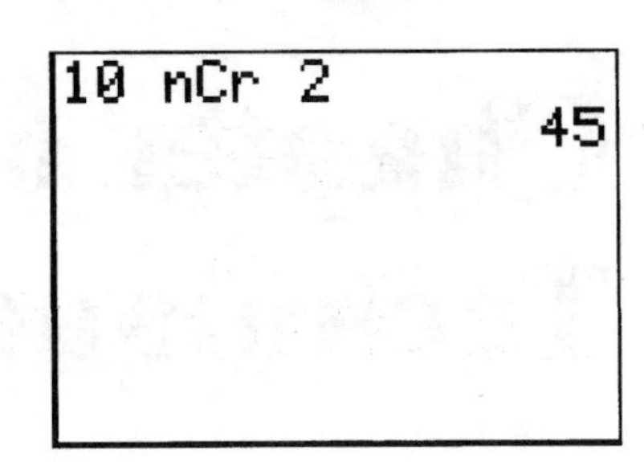

[1] [0] [MATH] [◄] [3] [2] [ENTER]

Thus, $C(10,2) = 45$.

63. How many house styles are possible if a contractor offers 3 choices of roof designs, 4 choices of window designs, and 6 choices of brick?

Apply the multiplication principle.

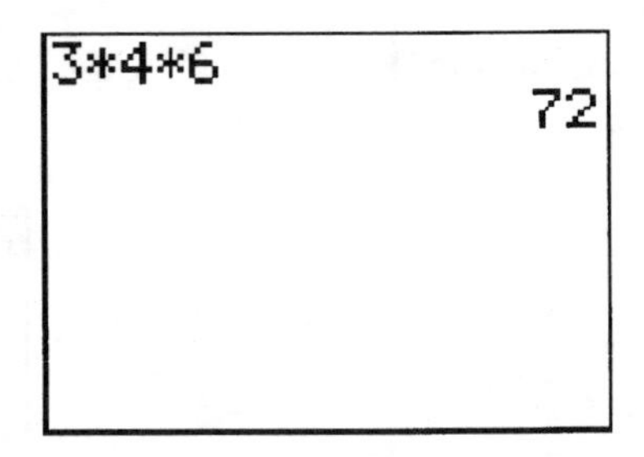

There are 72 different house styles.

65. How many different answers are possible in a true-false test consisting of 10 questions?

There are 2 possible answers for each question. Apply the multiplication principle.

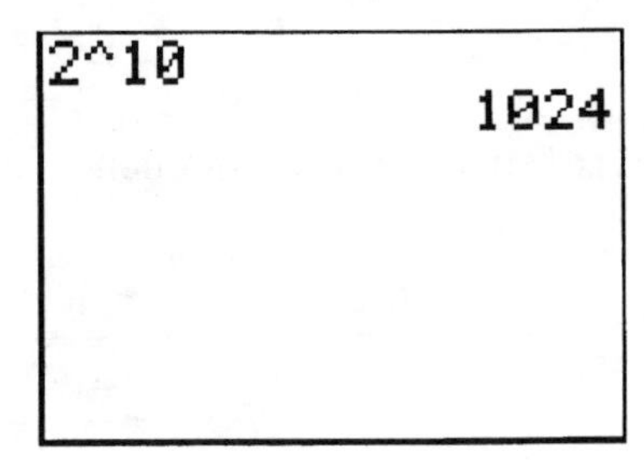

There are 1024 different ways to answer the 10 question test.

67. You are to set up a code of 3-digits words using the digits 1, 2, 3, 4, 5, 6 without using any digit more than once in the same word.

(a) What is the maximum number of words in such a language?

There are $P(6,3)$ possible words.

You can evaluate permutations using the nPr command found in the PRB sub menu of the [MATH] menu. The format is

n nPr *r*

[6] [MATH] [◄] [2] [3] [ENTER]

```
6 nPr 3
                120
```

Thus, $P(6,3)=120$, so there are 120 different words.

(b) If the words 124, 142, etc., designate the same word, how many different words are possible?

Since the order does not matter, now there are $C(6,3)$ different words.

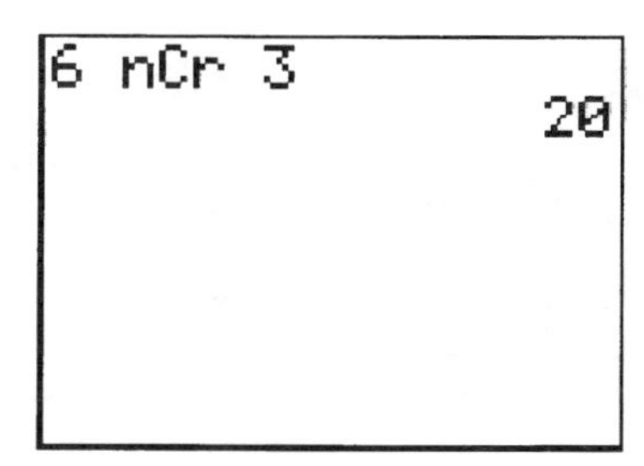

Thus, $C(6,3)=20$, so there are 20 different words.

71. **Arranging Books** A person has 4 history, 5 English, and 6 mathematics books. How many ways can the books be arranged on a shelf if books on the same subject must be together?

The total number of arrangements is given by $3!(4!5!6!)$.

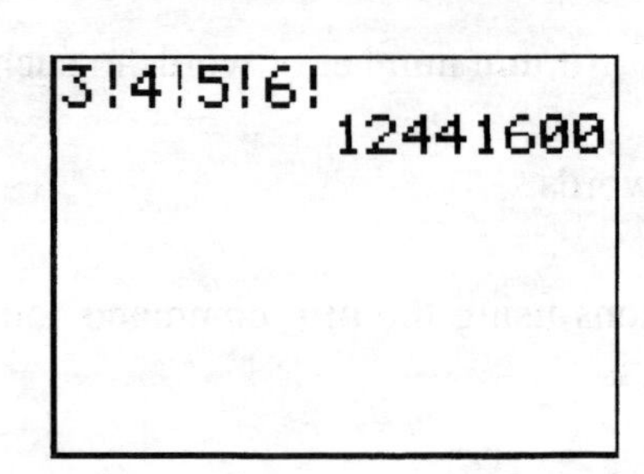

There are 12,441,600 different arrangements.

Summary

The commands introduced in this chapter are:

n `nCr` *r*

n `nPr` *r*

Chapter 7 – Probability

Section 7.1 Sample Spaces and the Assignment of Probabilities

Sometimes experiments are simulated using a random number function instead of actually performing the experiment. In Problems 57–62, use a graphing utility to simulate each experiment.

57. **Tossing a Fair Coin** Consider the experiment of tossing a fair coin. Simulate the experiment using a random number function, considering a toss to be tails (*T*) if the result is less than 0.5, and considering a toss to be heads (*H*) if the result is greater than or equal to 0.5. [*Note*: Most utilities repeat the action of the last entry if you simply press the ENTER, or EXE, key again.] Repeat the experiment 10 times. Using these 10 outcomes of the experiment you can estimate the probabilities $P(H)$ and $P(T)$ by the ratios

$$P(H) \approx \frac{\text{Number of times } H \text{ occured}}{10}$$

$$P(T) \approx \frac{\text{Number of times } T \text{ occured}}{10}$$

What are the actual probabilities? How close are the results of the experiment to the actual values?

To generate a random number between 0 and 1 on your calculator, use the `rand` function. This function can be found under the `PRB` submenu of the `MATH` menu.

[MATH] [◄] [1] [ENTER]

```
rand
          .7351965039
```

Since $0.7351965039 > 0.5$, our result is a heads.

NOTE: We probably won't have the same number. The random generator generates random numbers, these will vary from person to person.

To generate the next random number, press [ENTER] to execute the previous command, which was `rand`, again.

[ENTER]

```
rand
             .7351965039
              .351189551
```

Since $0.351189551 < 0.5$, our result is a tails. Generate 8 more random numbers. Keep track of all the random numbers.

[ENTER] [ENTER] [ENTER] [ENTER]

```
rand
             .7351965039
              .351189551
             .7291862237
             .4036456331
             .5404755792
              .138305685
```

[ENTER] [ENTER] [ENTER] [ENTER]

```
             .4036456331
             .5404755792
              .138305685
             .1535693143
             .7931370731
             .4270032014
             .5937218112
```

Count the number of values that are less than 0.5 and the number of values greater than or equal to 0.5. In this case, there are 5 values less than 0.5, and the remaining 5 values are greater than or equal to 0.5. Thus,

$$P(H) \approx \frac{5}{10} = 0.5$$

$$P(T) \approx \frac{5}{10} = 0.5$$

The actual probabilities are:

$$P(H) = \tfrac{1}{2} = 0.5$$

$$P(T) = \tfrac{1}{2} = 0.5$$

These results are identical to the exact values. Your results may be different because of the random number generator.

59. **Tossing a Loaded Coin** Consider the experiment of tossing a loaded coin. Simulate the experiment using a random number function, considering a toss to be tails (*T*) if the result is less than 0.25, and considering a toss to be heads (*H*) if the result is greater than or equal to 0.25. [*Note*: Most utilities repeat the action of the last entry if you simply press the ENTER, or EXE, key again.] Repeat the experiment 20 times. Using these 20 outcomes of the experiment you can estimate the probabilities $P(H)$ and $P(T)$ by the ratios

$$P(H) \approx \frac{\text{Number of times } H \text{ occured}}{20}$$

$$P(T) \approx \frac{\text{Number of times } T \text{ occured}}{20}$$

What are the actual probabilities? How close are the results of the experiment to the actual values?

Rather than generate random numbers one at a time, you can generate a list of random numbers using the `rand` command. If you wish to generate *n* random numbers, the format for the command is

`rand(`*n*`)`

Simulate tossing a coin 20 times (that is, generate 20 random numbers between 0 and 1). Store the resulting list in `L1`.

[MATH] [◄] [1] [(] [2] [0] [)] [STO▶] [2nd] [1] [ENTER]

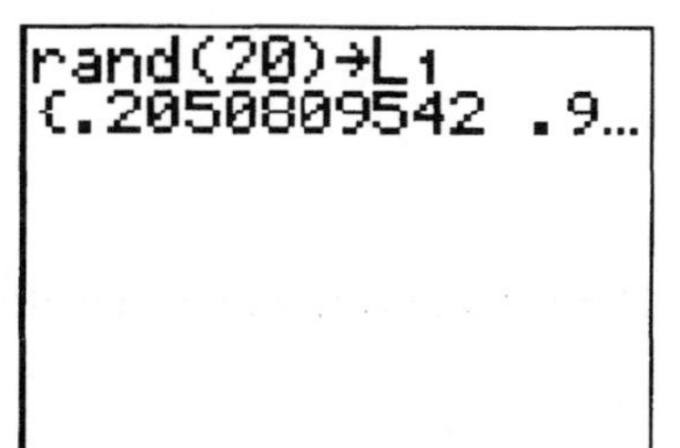

The 20 random numbers are stored in L1. To view the list, go the data editor in [STAT].

[STAT] [1]

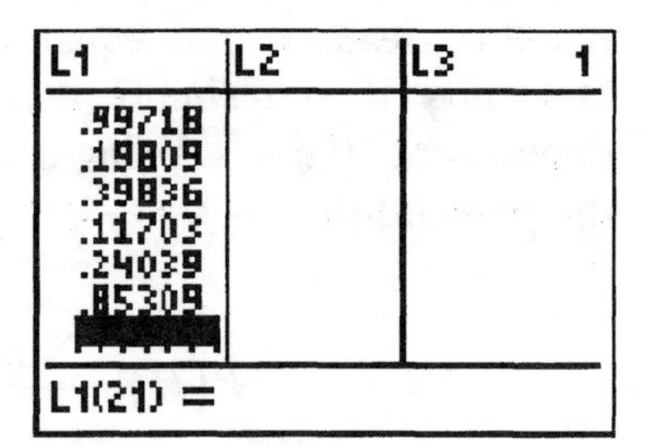

Use the [▼] and [▲] keys to scroll through the data. Note that the top three entries shown in the table below are the same three entries from the table shown above.

L1	L2	L3 1
.04279		
.11385		
.69654		
.98541		
.63263		
.31233		
.08574		

L1(14) =.085737147...

L1	L2	L3 1
.99718		
.19809		
.39836		
.11703		
.24039		
.85309		

L1(21) =

NOTE: We probably won't have the same 20 numbers. The random generator generates random numbers, these will vary from person to person.

Count the number of values that are less than 0.25 and the number of values greater than or equal to 0.25. In this case, there are 8 values less than 0.25, and the remaining 12 values are greater than or equal to 0.25. Thus,

$$P(H) \approx \frac{12}{20} = 0.6$$

$$P(T) \approx \frac{8}{20} = 0.4$$

The actual probabilities are:

$$P(H) = \tfrac{3}{4} = 0.75$$

$$P(T) = \tfrac{1}{4} = 0.25$$

These results are somewhat close, the probabilities are each off by 0.15.

61. **Jar and Marbles** Consider an experiment of choosing a marble from a jar containing 5 red, 2 yellow, and 8 white marbles. Simulate the experiment using a random number function on your calculator, considering a selection to be a red marble (R) if the result is less than or equal to 0.33, a yellow marble (Y) if the result is between 0.33 and 0.47, and a white marble (W) if the result is greater than or equal to 0.47. [*Note*: Most utilities repeat the action of the last entry if you simply press the ENTER, or EXE, key again.] Repeat the experiment 10 times. Using these 10 outcomes of the experiment, you can estimate the probabilities $P(R)$, $P(Y)$ and $P(W)$ by the ratios

$$P(R) \approx \frac{\text{Number of times } R \text{ occured}}{10}$$

$$P(Y) \approx \frac{\text{Number of times } Y \text{ occured}}{10}$$

$$P(W) \approx \frac{\text{Number of times } W \text{ occured}}{10}$$

What are the actual probabilities? How close are the results of the experiment to the actual values?

Simulate tossing a coin 10 times (that is, generate 10 random numbers between 0 and 1). Store the resulting list in L1.

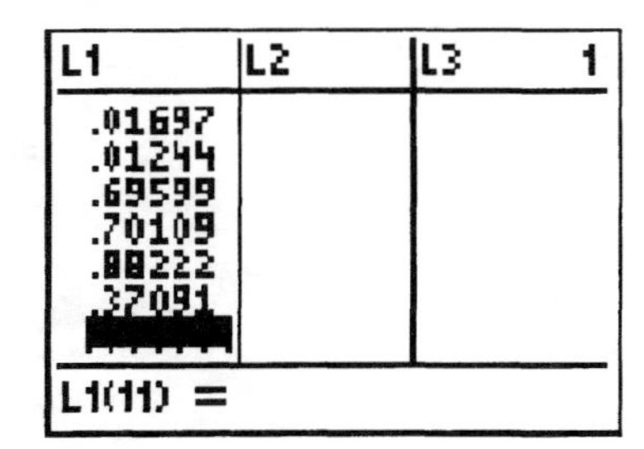

Go the data editor in STAT to view the 10 results.

L1	L2	L3 1
.37361	------	------
.59729		
.60633		
.07718		
.01697		
.01244		
.69599		

L1(1)=.3736142097...

L1	L2	L3 1
.01697		
.01244		
.69599		
.70109		
.88222		
.37091		

L1(11) =

Note that the top three entries shown in the table on the right are the same three entries from the table shown on the left.

NOTE: We probably won't have the same 10 numbers. The random generator generates random numbers, these will vary from person to person.

Count the number of values that are less than or equal to 0.33, the number of values between 0.33 and 0.47, and the number of values greater than or equal to 0.47. In this case, there are 4 values less than 0.33, 2 values that are between 0.33 and 0.47, and the remaining 4 values are greater than or equal to 0.47. Thus,

$$P(R) \approx \frac{4}{10} = 0.4$$

$$P(Y) \approx \frac{2}{10} = 0.2$$

$$P(W) \approx \frac{4}{10} = 0.4$$

The actual probabilities are:

$$P(R) = \tfrac{5}{15} = 0.333...$$

$$P(Y) = \tfrac{2}{15} = 0.133...$$

$$P(W) = \tfrac{8}{15} = 0.533...$$

These results are somewhat close, two of the estimates are within 0.0666… of the actual probability, while the other estimate is off by 0.1333….

Summary

The commands introduced in this chapter are:

```
rand

rand(n)
```

Chapter 8 – Additional Probability Topics

Section 8.2 The Binomial Probability Model

Sometimes experiments are simulated using a random number function instead of actually performing the experiment.

In Problems 47 – 50, use a graphing utility to simulate each experiment.

47. **Tossing a Fair Coin** Consider the experiment of tossing a fair coin 4 times, counting the number of heads occurring in these 4 tosses. Simulate the experiment using a random number function on your calculator, considering a toss to be tails (*T*) if the result is less than 0.5, and considering a toss to be heads (*H*) if the result is greater than or equal to 0.5. Record the number of heads in 4 tosses. [*Note*: Most calculators repeat the action of the last entry if you simply press the ENTER, or EXE, key again.] Repeat the experiment 10 times, obtaining a sequence of 10 numbers. Using these 10 numbers you can estimate the probability of *k* heads, $P(k)$, for each $k = 0, 1, 2, 3, 4$ by the ratio

$$\frac{\text{Number of times } k \text{ appears in your sequence}}{10}$$

Enter your estimates in the table below. Calculate the actual probabilities using the binomial probability formula, and enter these numbers in the table. How close are your numbers to the actual values?

k	Your Estimate of $P(k)$	Actual Value of $P(k)$
0		
1		
2		
3		
4		

Recall from Chapter 7, we can generate random numbers using the `rand` command. To simulate tossing the coin four times, use `rand(4)`, store the results in the list L1, and go to the data editor under [STAT] to view the results. Remember that we are using a random number generator, so your results will most likely differ from the results shown on the next page.

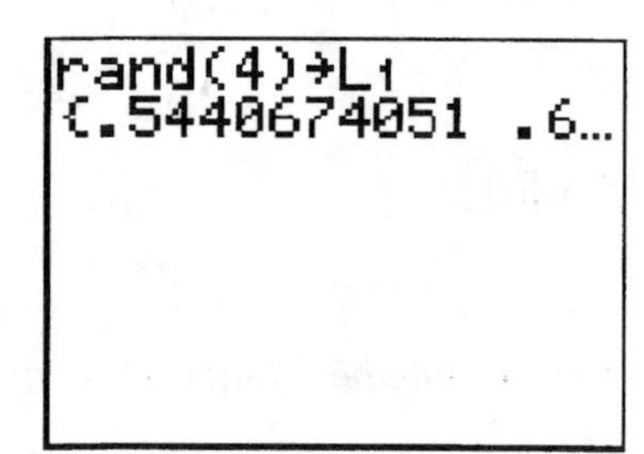

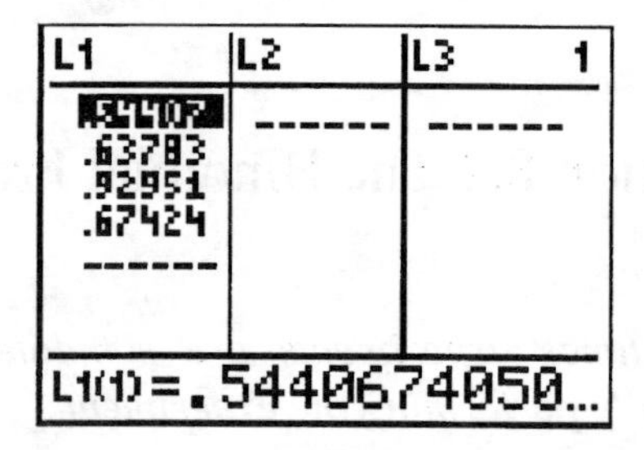

All 4 entries are greater than 0.5, so our result for our first trial is 4. Return to the home screen and press [ENTER] to repeat the experiment again.

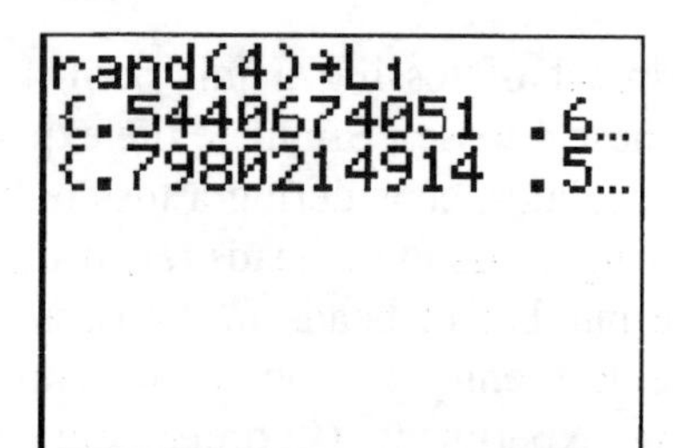

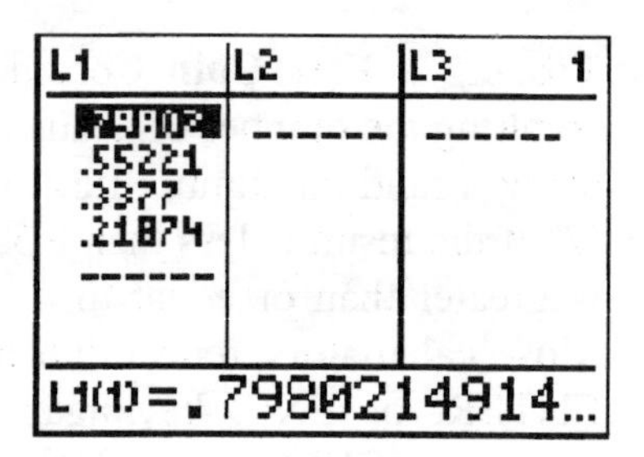

This time there were only two entries greater than 0.5, so the result from our second trial is 2. Repeat the experiment again.

This time there were only two entries greater than 0.5, so the result from our third trial is 2. Repeat the experiment again.

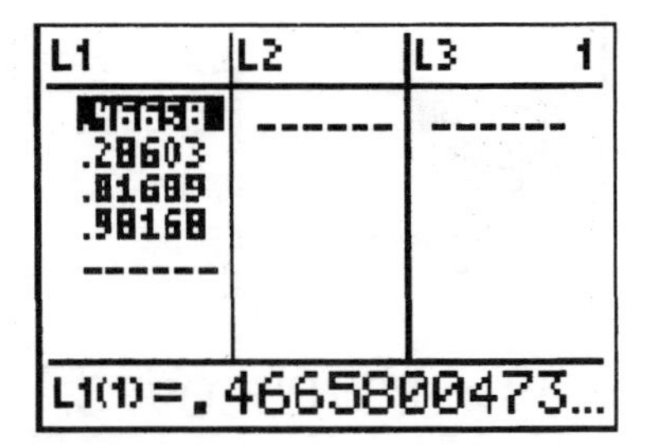

This time there were two entries greater than 0.5, so the result from our fourth trial is 2. Repeat the experiment again.

L1 L2 L3 1
.4042
.2291
.6279
L1(1)=.4919710165...

This time there was only one entry greater than 0.5, so the result from our fifth trial is 1. Repeat the experiment again.

L1 L2 L3 1
.80684
.77343
.46384
L1(1)=.6174036362...

This time there were three entries greater than 0.5, so the result from our sixth trial is 3. Repeat the experiment again.

L1 L2 L3 1
.62438
.33353
.45469
L1(1)=.9251782617...

This time there were two entries greater than 0.5, so the result from our seventh trial is 2. Repeat the experiment again.

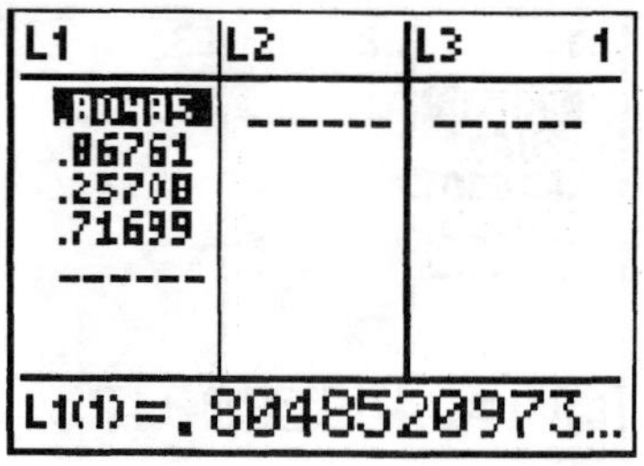

```
L1      L2      L3     1
.80485  ------  ------
.86761
.25708
.71699
------

L1(1)=.8048520973...
```

This time there were three entries greater than 0.5, so the result from our eighth trial is 3. Repeat the experiment again.

```
L1      L2      L3     1
.46813  ------  ------
.4772
.21292
.09479
------

L1(1)=.4681336964...
```

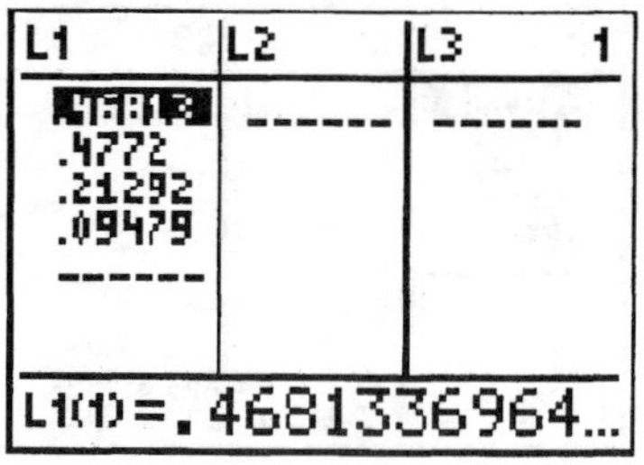

This time there were no entries greater than 0.5, so the result from our ninth trial is 0. Repeat the experiment again.

```
L1      L2      L3     1
.3614   ------  ------
.63473
.13745
.66654
------

L1(1)=.3613961004...
```

This time there were two entries greater than 0.5, so the result from our tenth trial is 2.

The sequence of results obtained is $\{4, 2, 2, 2, 1, 3, 2, 3, 0, 2\}$. Using these results, we estimate the probabilities $P(k)$, for each $k = 0, 1, 2, 3, 4$. Enter these estimates, as well as the exact values, in the table below.

k	Your Estimate of $P(k)$	Actual Value of $P(k)$
0	$\frac{1}{10} = 0.1$	$\frac{1}{16} = 0.0625$
1	$\frac{1}{10} = 0.1$	$\frac{4}{16} = 0.25$
2	$\frac{5}{10} = 0.5$	$\frac{6}{16} = 0.375$
3	$\frac{2}{10} = 0.2$	$\frac{4}{16} = 0.25$
4	$\frac{1}{10} = 0.1$	$\frac{1}{16} = 0.0625$

While the estimates are not real close to the actual values, the biggest difference between actual and estimate is only 0.15.

49. **Tossing a Fair Coin** Consider the experiment of tossing a fair coin 8 times, counting the number of heads occurring in these 8 tosses. Simulate the experiment using a random number function on your calculator, considering a toss to be tails (*T*) if the result is less than 0.5, and considering a toss to be heads (*H*) if the result is greater than or equal to 0.5. Record the number of heads in 8 tosses. Repeat the experiment 10 times, obtaining a sequence of 10 numbers. Using these 10 numbers you can estimate the probability of *3* heads, $P(3)$, by the ratio

$$\frac{\text{Number of times 3 appears in your sequence}}{10}$$

Calculate the actual probability using the binomial probability formula. How close is your estimate to the actual value?

To simulate tossing the coin eight times, use `rand(8)`. Store the results in the list L1, and go to the data editor under STAT to view the results. Remember that we are using a random number generator, so your results will most likely differ from the results shown on the next page.

```
rand(8)→L1
{.0125324079 .5...
```

Be sure to scroll down the list to see all the entries.

```
L1      L2      L3     1
.01253  ------  ------
.59775
.27714
.01111
.92064
.83589
.25534
L1(1)=.0125324079...
```

```
L1      L2      L3     1
.27714
.01111
.92064
.83589
.25534
.72567
------
L1(8)=.7256726824...
```

There are 4 entries are greater than 0.5, so our result for our first trial is 4. Return to the home screen and press ENTER to repeat the experiment again.

This time there four entries greater than 0.5, so the result from our second trial is 4. Repeat the experiment again.

This time there were only three entries greater than 0.5, so the result from our third trial is 3. Repeat the experiment again.

This time there were two entries greater than 0.5, so the result from our fourth trial is 2. Repeat the experiment again.

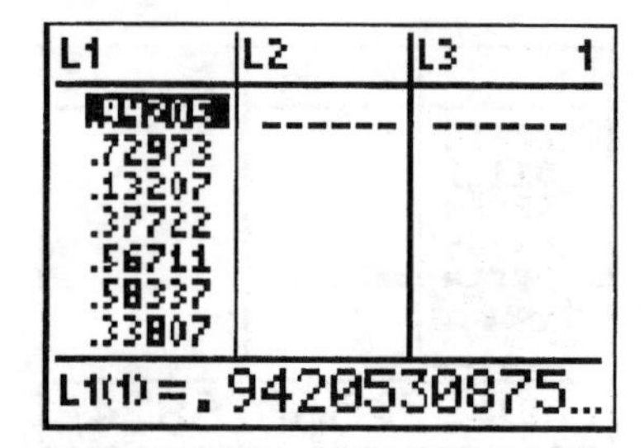

This time there four entries greater than 0.5, so the result from our fifth trial is 4. Repeat the experiment again.

L1 L2 L3 1
.89735
.19028
.52179
.3266
.70004
.96285
L1(1)=.2795177400...

L1 L2 L3 1
.19028
.52179
.3266
.70004
L1(8)=.9246107989...

This time there were five entries greater than 0.5, so the result from our sixth trial is 5. Repeat the experiment again.

L1 L2 L3 1
.69766
.01518
.10182
.79163
.36884
.50691
L1(1)=.6485679702...

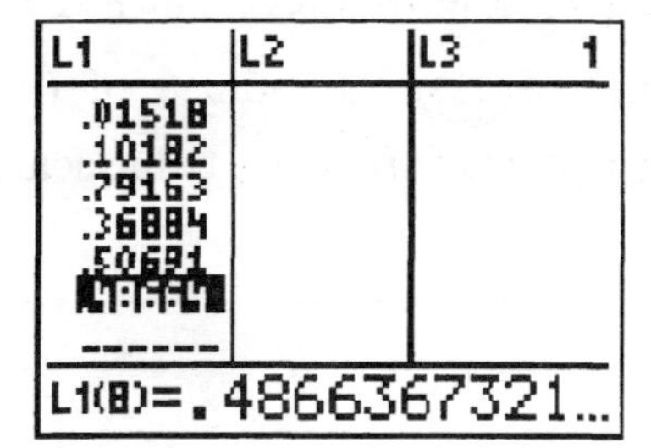

This time there were four entries greater than 0.5, so the result from our seventh trial is 4. Repeat the experiment again.

L1 L2 L3 1
.39844
.17474
.43161
.9881
.23712
.48703
L1(1)=.9172764946...

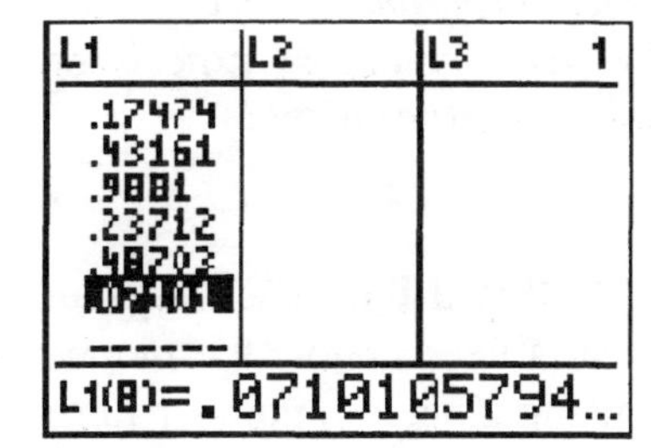

This time there were two entries greater than 0.5, so the result from our eighth trial is 2. Repeat the experiment again.

L1 L2 L3 1
.31665
.50138
.40502
.23232
.10583
.96065
L1(1)=.9720610387...

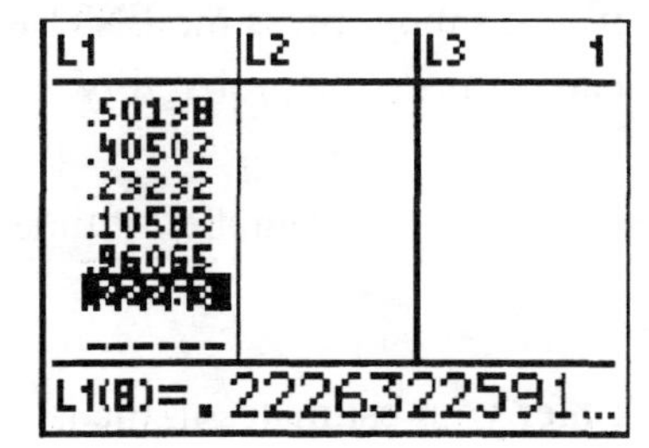

This time there were three entries greater than 0.5, so the result from our ninth trial is 3. Repeat the experiment again.

This time there were two entries greater than 0.5, so the result from our tenth trial is 2.

The sequence of results obtained is $\{4, 4, 3, 2, 4, 5, 4, 2, 3, 2\}$. There are two occurrences of 3 in the list, so the estimate of $P(3)$ **is** $\frac{2}{10} = 0.2$. The exact value is $P(3) = 0.21875$. The estimate is fairly close to the actual value.

Section 8.5 Random Variables

Sometimes experiments are simulated using a random number function instead of actually performing the experiment. In Problems 9 – 14, use a graphing utility to simulate each experiment.

9. **Rolling a Fair Die** Consider the experiment of rolling a die, and let the random variable X denote the number showing on the top face. Simulate the experiment using a random number function on your calculator, considering a roll to have the outcome k if the value of the random number function is between $(k-1)\cdot 0.167$ and $k\cdot 0.167$. Record the outcome. Repeat the experiment 50 times, obtaining a sequence of 50 numbers. [*Note*: Most calculators repeat the action of the last entry if you simply press the ENTER, or EXE, key again.] Using these 50 numbers you can estimate the probability $P(X = k)$ for $k = 0, 1, 2, 3, 4, 5, 6$ by the ratio

$$\frac{\text{Number of times } k \text{ appears in your sequence}}{50}$$

Enter your estimates in the table. Calculate the actual probabilities and enter these numbers in the table. How close are your numbers to the actual values?

k	Your Estimate of $P(X=k)$	Actual Value of $P(X=k)$
0		
1		
2		
3		
4		
5		
6		

Recall from Chapter 7, we can generate random numbers using the `rand` command. To simulate rolling a die fifty times, use `rand(5Ø)`. Store the results in the list L1, and go to the data editor under STAT to view the results. Remember that we are using a random number generator, so your results will most likely differ from the results shown on the next page.

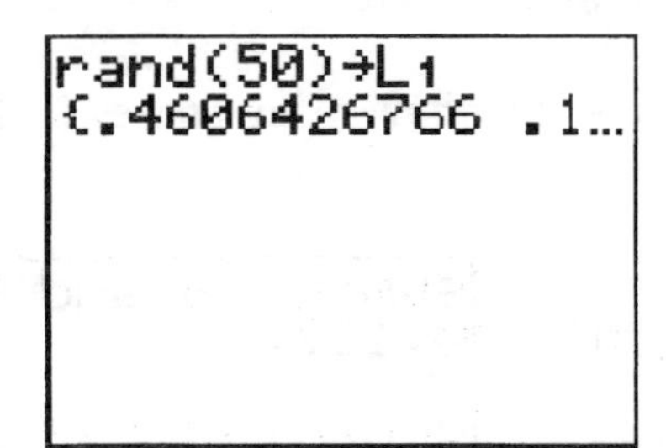


```
L1       L2      L3     1
.46064   ------  ------
.14467
.9469
.83103
.52085
.09735
.7925
L1(1)=.4606426766...
```

With 50 entries in the list, it can be difficult to count how many entries correspond to rolling a 1, rolling a 2, etc. You can have your calculator count how many entries match rolling a 1 (that is, count how many numbers were between 0 and 0.167), how many entries match rolling a 2 (that is, count how many numbers were between 0 and 0.334), etc.

We can do this using the `sum` command. The sum command is found in the MATH submenu of the `[LIST]` menu. The format for the command is

`Sum(`*list*`)`

To count the number of values in a list that fall in a desired interval, say between 0 and 0.167 for rolling a 1, we test the values in the list using inequalities. To count the values in a list that are between 0 and 0.167 we would use

```
Sum(L1>Ø and L1<Ø.167)
```

Remember that an inequality of the form $a<x<b$ is equivalent to $a<x$ **and** $x<b$.

The inequality symbols are found under the [TEST] menu and the logical operator and is found under the LOGIC submenu of [TEST].

Be sure to exit the data editor first.

Determine the number of times a 1 was rolled. That is, count how many entries in the list are between 0 and 0.167.

[2nd] [STAT] [◄] [5] [2nd] [1] [2nd] [MATH] [3] [0] [2nd] [MATH] [►]

[1] [2nd] [1] [2nd] [MATH] [5] [0] [.] [1] [6] [7] [)] [ENTER]

sum(L1>0 and L1<
0.167)
5

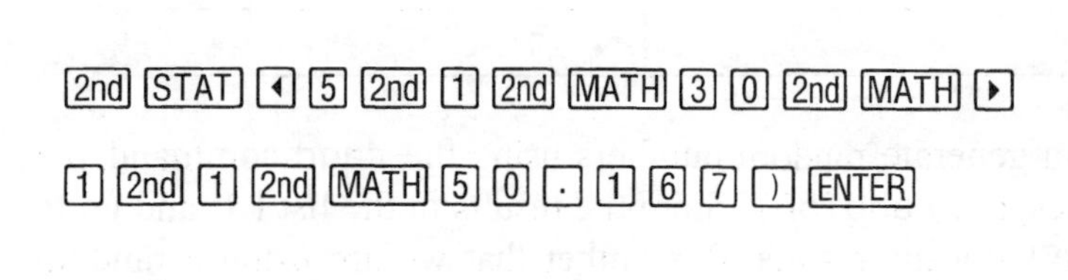

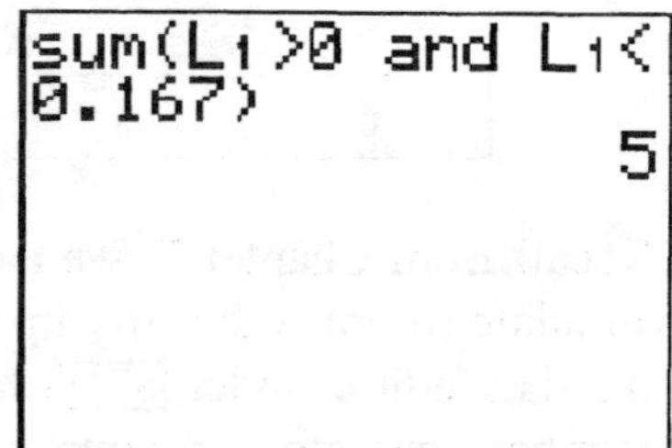

There were 5 numbers between 0 and 0.167, so there were 5 1's rolled.

Determine the number of times a 2 was rolled by counting how many entries in the list are between 0.167 and 0.334.

[2nd] [STAT] [◄] [5] [2nd] [1] [2nd] [MATH] [3] [0] [.] [1] [6] [7]

[2nd] [MATH] [►] [1] [2nd] [1] [2nd] [MATH] [5] [0] [.] [3] [3] [4] [)]

[ENTER]

sum(L1>0 and L1<
0.167)
5
sum(L1>0.167 and
L1<0.334)
7

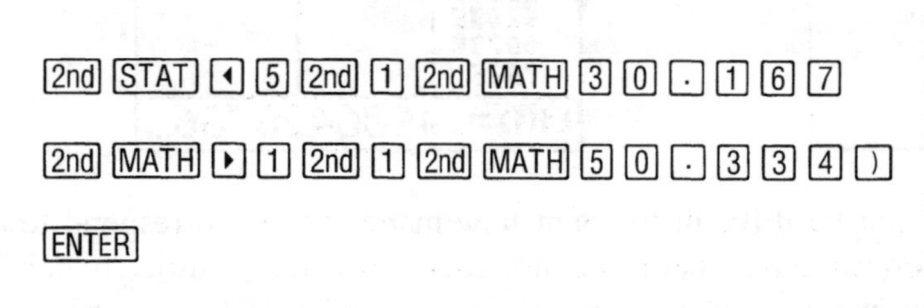

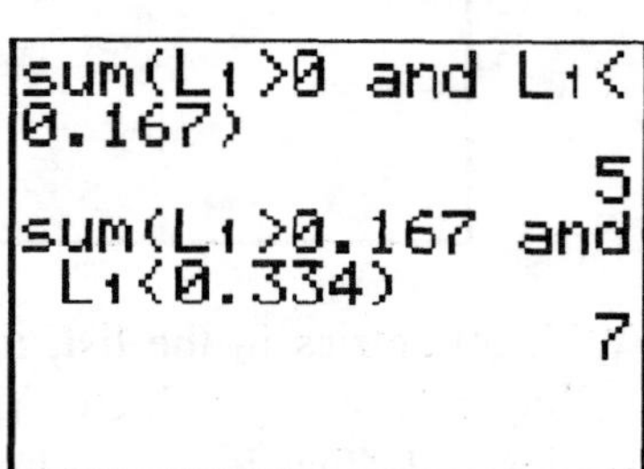

There were 7 numbers between 0.167 and 0.334, so there were 7 2's rolled.

Determine the number of times a 3 was rolled by counting how many entries in the list are between 0.334 and 0.501. Then determine the number of times a 4 was rolled by counting how many entries in the list are between 0.501 and 0.668.

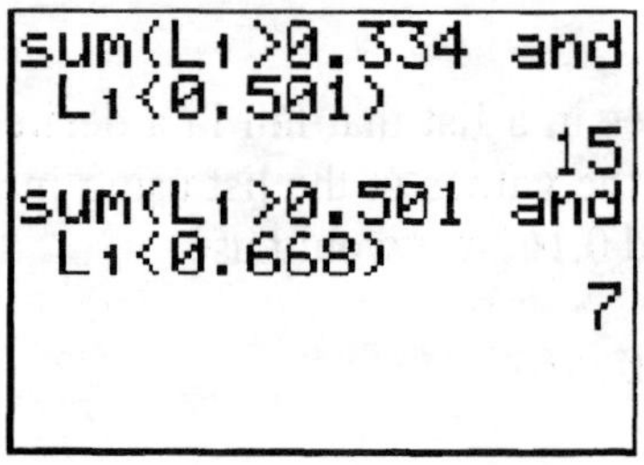

sum(L1>0.334 and
L1<0.501)
15
sum(L1>0.501 and
L1<0.668)
7

There were 15 numbers between 0.334 and 0.501, so there were 15 3's rolled. There were 7 numbers between 0.501 and 0.668, so there were 7 4's rolled.

Determine the number of times a 5 was rolled by counting how many entries in the list are between 0.668 and 0.835. Then determine the number of times a 6 was rolled by counting how many entries in the list are between 0.835 and 1.

```
sum(L1>0.668 and
 L1<0.835)
                8
sum(L1>0.835 and
 L1<1)
                8
```

There were 8 numbers between 0.668 and 0.835, so there were 8 5's rolled. There were 8 numbers between 0.835 and 1, so there were 8 6's rolled.

In the table below, enter the ratios using that data from above. Calculate the actual probabilities and enter those in the table as well.

k	Your Estimate of $P(X = k)$	Actual Value of $P(X = k)$
1	$\frac{5}{50} = 0.1$	$\frac{1}{6} \approx 0.167$
2	$\frac{7}{50} = 0.14$	$\frac{1}{6} \approx 0.167$
3	$\frac{15}{50} = 0.3$	$\frac{1}{6} \approx 0.167$
4	$\frac{7}{50} = 0.14$	$\frac{1}{6} \approx 0.167$
5	$\frac{8}{50} = 0.16$	$\frac{1}{6} \approx 0.167$
6	$\frac{8}{50} = 0.16$	$\frac{1}{6} \approx 0.167$

Most of the estimates are close to the exact value, but one estimate is different than the exact value by 0.133.

11. Use a random number function to select a value for the random variable X. Repeat this experiment 50 times. Count the number of times the random variable X is between 0.1 and 0.3. Calculate the ratio

$$R = \frac{\text{Number of times the random variable } X \text{ is between 0.1 and 0.3}}{50}$$

What value of R did you obtain? Calculate the actual probability $P(0.1 \le X \le 0.3)$.

Using the procedure discussed in Problem 9 of this section, we can generate a list of 50 random numbers, and we can easily find out how many are between 0.1 and 0.3. Remember that we are using a random number generator, so your results will most likely differ from the results shown below.

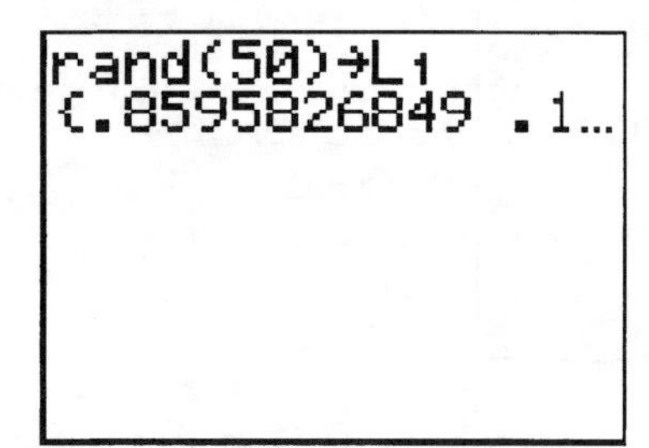

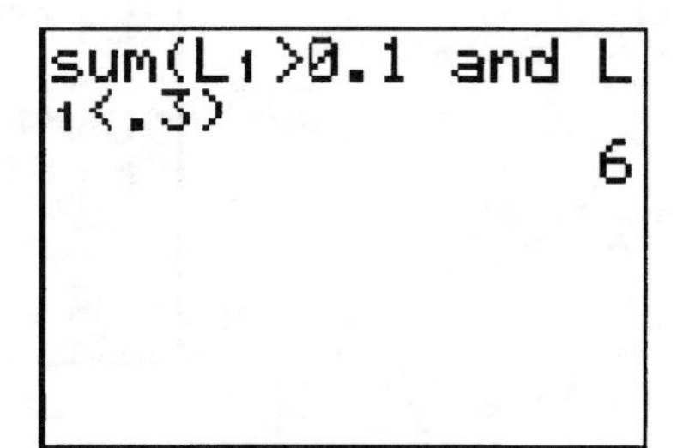

There are 6 entries that are between 0.1 and 0.3. Thus, $R = \frac{6}{50} = 0.12$.

The exact value is $P(0.1 \le X \le 0.3) = 0.2$. The estimate is somewhat close to the exact, the difference between the two values is 0.08.

13. **Rolling a Dodecahedron** Consider the experiment of rolling a dodecahedron (a regular polyhedron with all 12 faces congruent), and let the random variable X denote the number showing on the top face. Simulate the experiment using a random number function on your calculator, considering a roll to have the outcome k if the value of the random number function is between $(k-1)/12$ and $k/12$ for $k = 1, 2, 3, 4, 5, \ldots, 12$. Record the outcome. Repeat the experiment 50 times, thus obtaining a sequence of 50 numbers. Using the 50 numbers you can estimate the probability $P(X = 2)$ by the ratio

$$\frac{\text{Number of times 2 appears in your sequence}}{50}$$

Calculate the actual probability, $P(X = 5)$, and compare these values. How close is your estimate to the actual value?

Using the procedure discussed in Problem 9 of this section, we can generate a list of 50 random numbers, and we can easily find out how many are between $1/12$ and $2/12$.

Remember that we are using a random number generator, so your results will most likely differ from the results shown below.

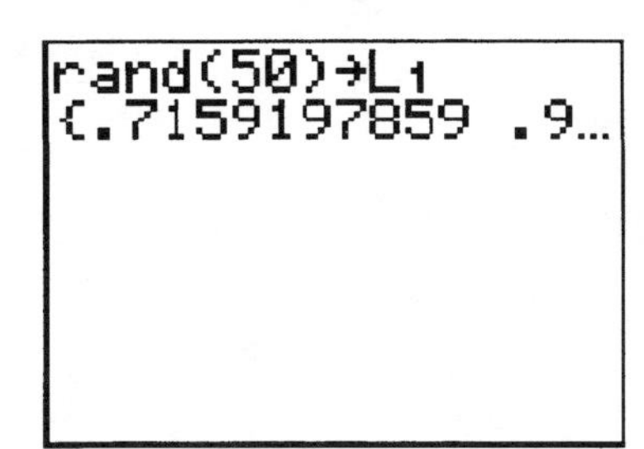

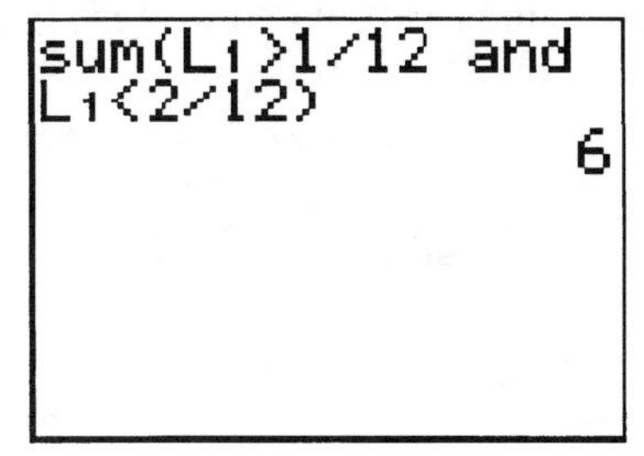

There are 9 entries that are between $1/12$ and $2/12$. Thus, $P(X=2)\approx \frac{6}{50}=0.12$.

The exact value is $P(X=2)=\frac{1}{12}=0.08333\ldots$. The estimate is somewhat close to the exact, the difference between the two values is $0.0366\ldots$.

Summary

The commands introduced in this chapter are:

`Sum(`*list*`)`

`and`

`<`

Chapter 9 – Statistics

Section 9.6 The Normal Distribution

37. Graph the standard normal curve using a graphing utility. For what value of x does the function assume its maximum? The equation is given by

$$y = \frac{1}{\sqrt{2\pi}} e^{-(1/2)x^2}$$

To graph a function, remember you must first enter the equation in the function editor.

```
Plot1 Plot2 Plot3
\Y1=1/√(2π)*e^(-
(1/2)X²)
\Y2=
\Y3=
\Y4=
\Y5=
\Y6=
```

Next, set an appropriate viewing window.

```
WINDOW
 Xmin=-5
 Xmax=5
 Xscl=1
 Ymin=0
 Ymax=.5
 Yscl=.1
 Xres=1
```

Graph the function.

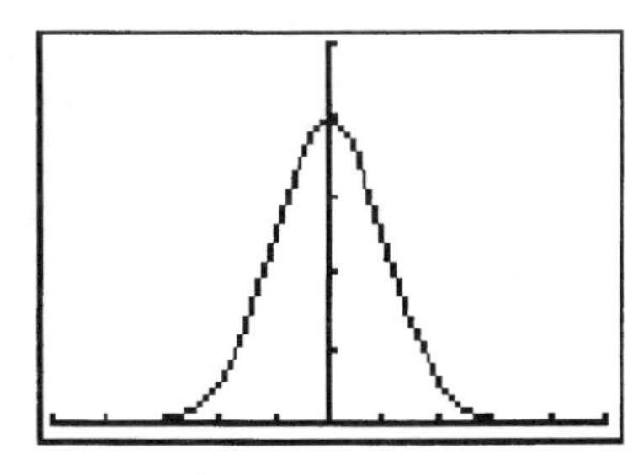

It appears that the maximum occurs when $x = 0$.

We can use the calculator to find the maximum using the `maximum` function, which can be found in the `CALC` menu. To find the maximum, we must give the calculator a left bound (an x-value that is less than the x-coordinate of the maximum), a right bound (an x-value that is greater than the x-coordinate of the maximum), and a guess for the x-coordinate of the maximum. You can enter values for each of the parameters the calculator requires.

For our problem, note that the x-value of the maximum is between $x = -1$ and $x = 1$, so we will use these as our bounds.

[2nd] [TRACE] [4] [(-)] [1]

Y1=1/√(2π)*e^(-(1/2)X²)
Left Bound?
X= -1

[ENTER] [1]

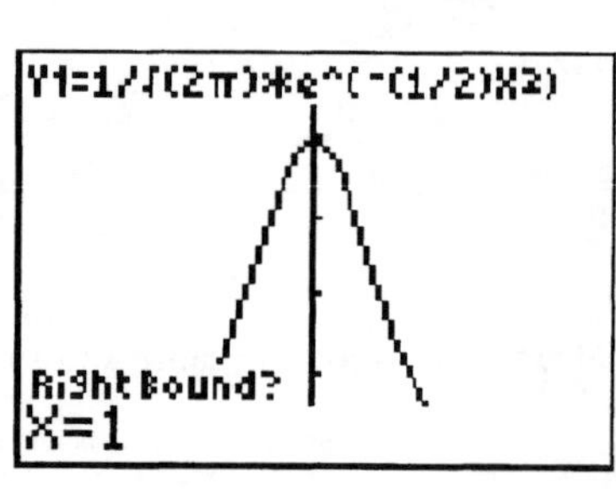

Use $x = 0$ as our guess for this problem.

[ENTER] [0]

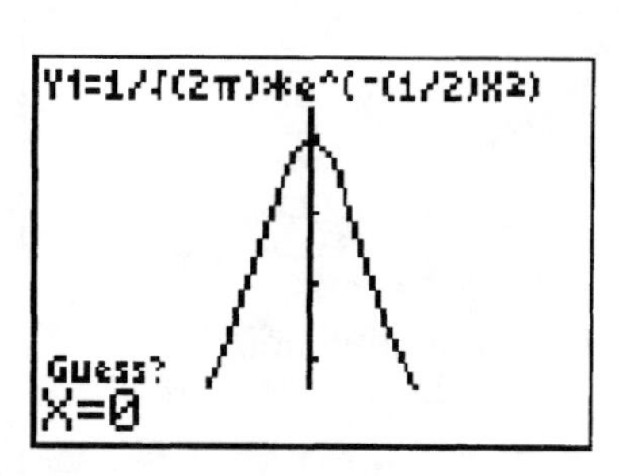

Find the maximum.

[ENTER]

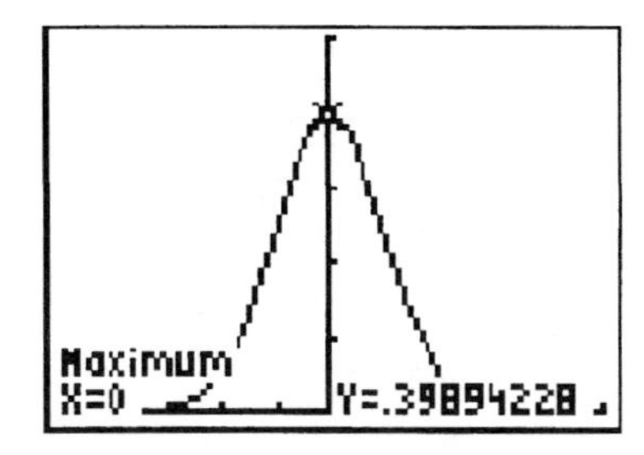

The maximum occurs when $x=0$ and the maximum value is $y \approx 0.39894228$. Note that if we substitute $x=0$ into the equation, we get the exact value of $y=\frac{1}{\sqrt{2\pi}}$.

39. **Quality Control** Refer to Problem 35. Use a graphing utility to compute the exact probability that the sample contains at least 10 bags of improperly sealed jelly beans.

While we can use the normal distribution to approximate the binomial distribution when the n is large, to calculate the exact probability, we must use the binomial distribution. To calculate cumulative probabilities using a binomial distribution, we use the `binomcdf(` function on the calculator. The `binomcdf(` function will find the probability of at most x successes out of n trials if the probability of success on any one trial is p. The format is

`binomcdf(`n, p, x`)`

You will find the `binomcdf` command under the `[DISTR]` menu.

Since we want the probability of at least 10 bags, we will need to use the complement of an event rule. That is,

$$P(\text{at least } 10) = 1 - P(\text{at most } 9)$$

[1] [-] [2nd] [VARS] [▲] [▲] [▲] [▲] [▲] [ENTER] [5] [0] [0] [,] [.]

[0] [1] [,] [9] [)] [ENTER]

```
1-binomcdf(500,.
01,9)
         .0311021071
```

Thus, $P(X \geq 10) \approx 0.0311$.

Summary

The commands introduced in this chapter are:

`maximum`

`binomcdf(`*n*, *p*, *x*`)`

Chapter 10 – Functions and Their Graphs

Section 10.1 Graphs of Equations

In Problems 11–22, the graph of an equation is given. (a) List the intercepts of the graph. (b) Based on the graph, tell whether the graph is symmetric with respect to the x-axis, the y-axis, and/or the origin.

19.

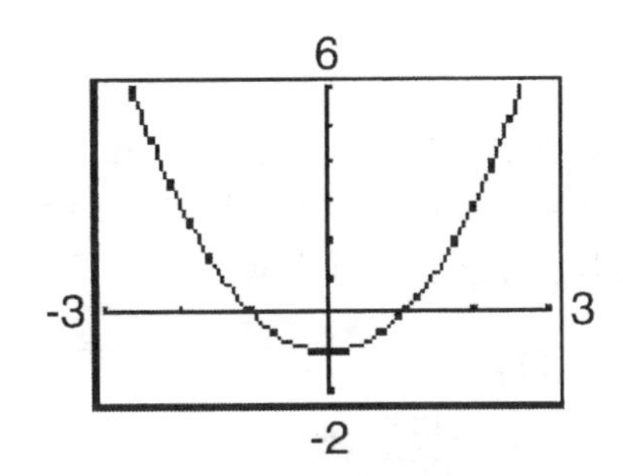

(a) The x-intercepts are $(1,0)$ and $(-1,0)$, and the y-intercept is $(0,-1)$.

(b) The graph is symmetric with respect to the y-axis.

21.

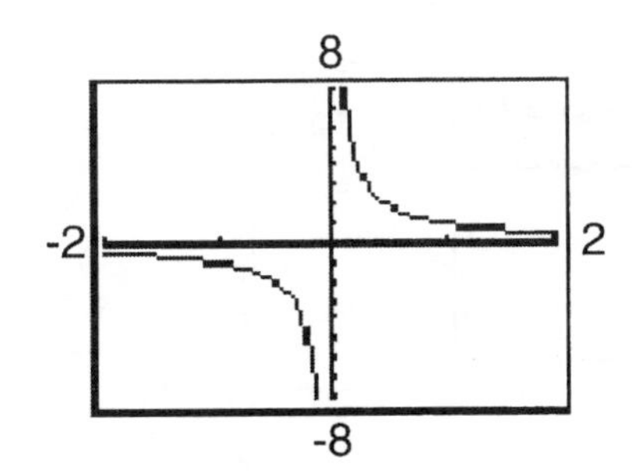

(a) This graph does not have any x-intercepts or y-intercepts.

(b) The graph is symmetric with respect to the origin.

Section 10.3 Graph of a Function; Properties of Functions

In Problems 69–76, use a graphing utility to graph each function over the indicated interval and approximate any local maxima and local minima. Determine where the function is increasing and where it is decreasing. Round answers to two decimal places.

69. $f(x) = x^3 - 3x + 2 \quad (-2, 2)$

Enter the formula for f in the function editor. Go to `WINDOW` and enter limits for x. We must determine limits for y, so try $-10 \le y \le 10$. If this does not work, then we can go back to `WINDOW` and adjust the limits on y until we find a good window.

```
Plot1 Plot2 Plot3
\Y1=X^3-3X+2
\Y2=
\Y3=
\Y4=
\Y5=
\Y6=
\Y7=
```

```
WINDOW
 Xmin=-2
 Xmax=2
 Xscl=1
 Ymin=-10
 Ymax=10
 Yscl=1
 Xres=1
```

Graph the function.

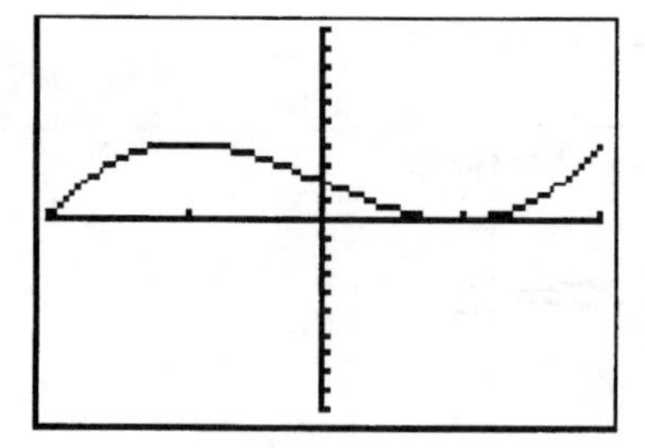

This window is suitable.

You can use your TI-83 Plus to find the x- and y-coordinates of any local maxima and/or local minima using the `maximum` or `minimum` functions respectively. Both the `maximum` and `minimum` functions require three inputs: an x-value to the left of the maxima or minima; an x-value to the right of the maxima or minima; and an estimate of the x-value of the maxima or minima. Both `maximum` and `minimum` can be found under the `[CALC]` menu.

Find the local maximum.

[2nd] [TRACE]

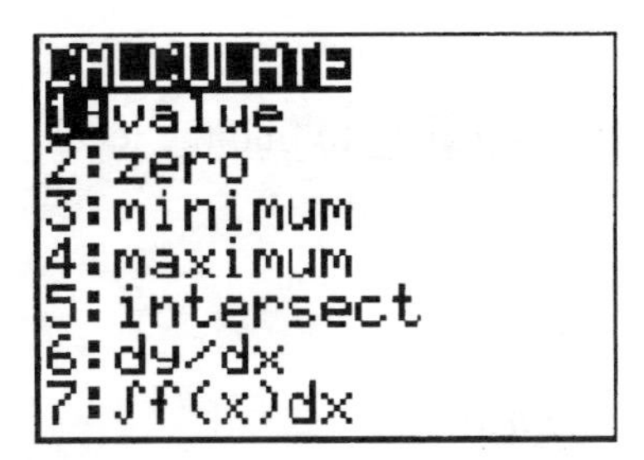

Select `maximum`. Enter a value for x that is less than (to the left of) the x–coordinate of the local maximum. Notice that the local maximum is between $x = -2$ and $x = 0$, so we can use $x = -2$ as a left bound.

[4] [(-)] [2]

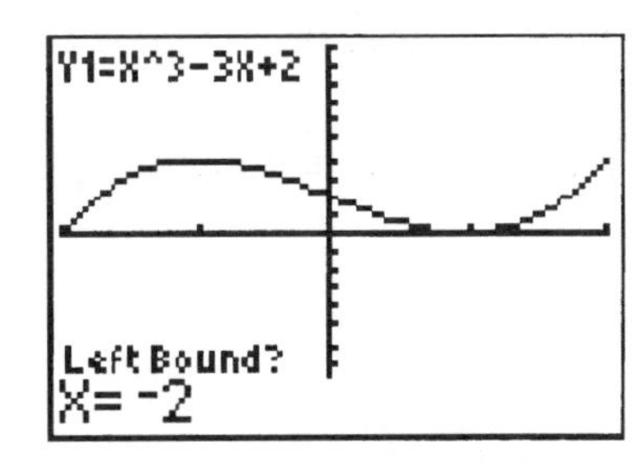

Input $x = 0$ as a right bound.

[ENTER] [0]

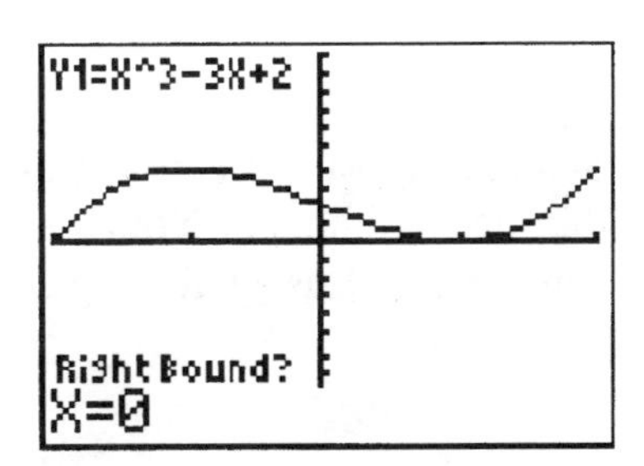

Input $x = -1$ as a guess.

[ENTER] [(-)] [1]

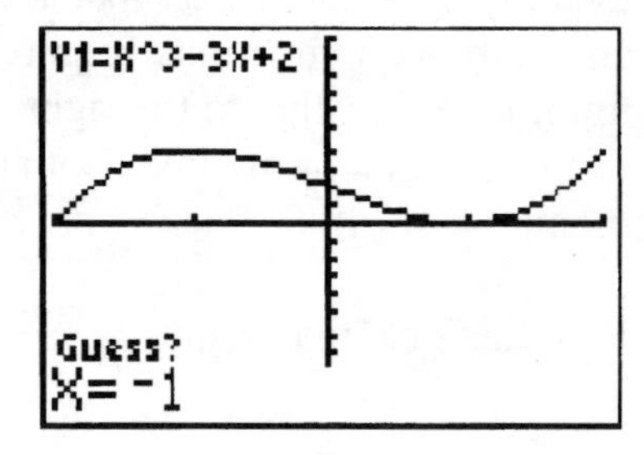

Press [ENTER] to find the local maximum.

[ENTER]

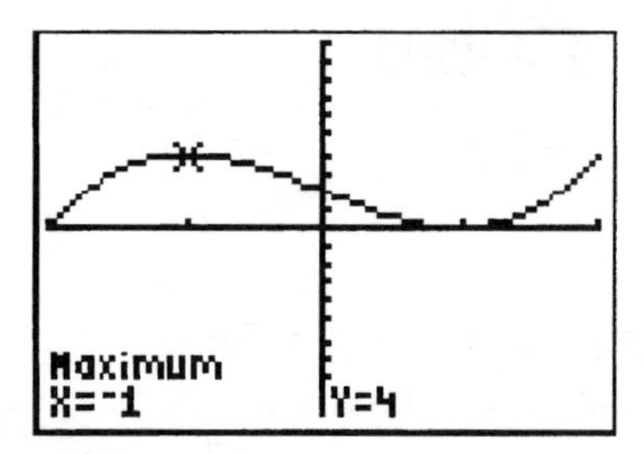

Thus the local maximum is $(-1, 4)$.

Find the local minimum.

[2nd] [TRACE]

Select `minimum`. Enter a value for x that is less than (to the left of) the x–coordinate of the local minimum. Notice that the local minimum is between $x = 0$ and $x = 2$, so we can use $x = 0$ as a left bound.

[3] [0]

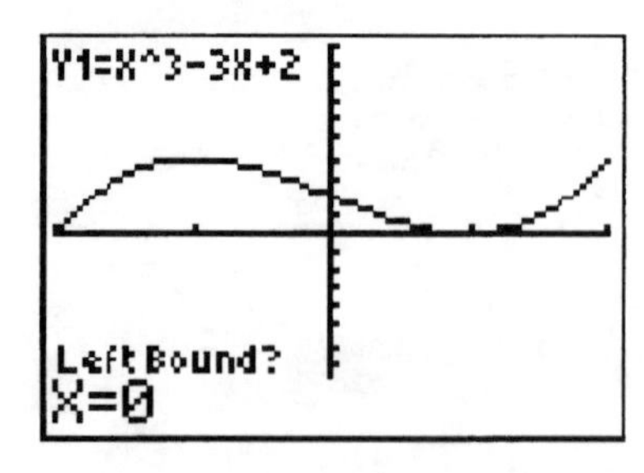

Input $x = 2$ as a right bound.

[ENTER] [2]

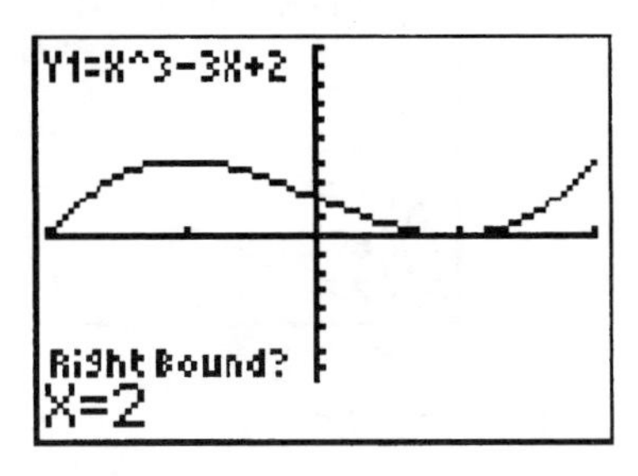

Input $x = 1$ as a guess.

[ENTER] [1]

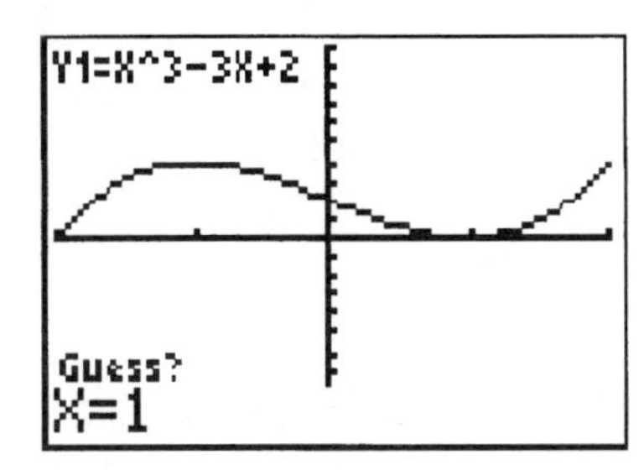

Press [ENTER] to find the local minimum.

[ENTER]

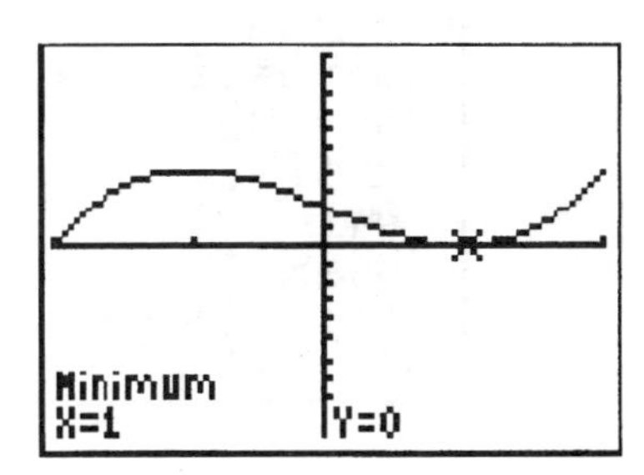

Thus the local minimum is $(1, 0)$.

Thus, f is increasing on the intervals $(-2,-1)$ and $(1,2)$, and f is decreasing on the interval $(-1,1)$.

71. $f(x) = x^5 - x^3 \quad (-2, 2)$

Enter the formula for f in the function editor. Go to WINDOW and enter limits for x. We must determine limits for y, so try $-10 \leq y \leq 10$. If this does not work, then we can go back to WINDOW and adjust the limits on y until we find a good window.

```
Plot1 Plot2 Plot3
\Y1=X^5-X^3
\Y2=
\Y3=
\Y4=
\Y5=
\Y6=
\Y7=
```

```
WINDOW
 Xmin=-2
 Xmax=2
 Xscl=1
 Ymin=-10
 Ymax=10
 Yscl=1
 Xres=1
```

Graph the function.

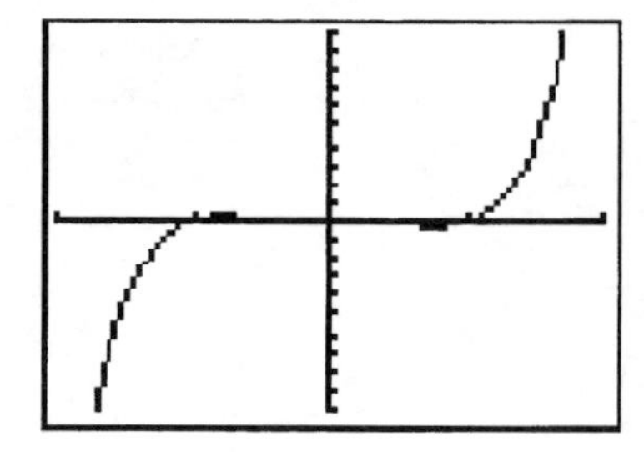

This window is not suitable; we cannot see the portion of the graph between $x = -1$ and $x = 1$. We need to reduce the limits for *y* so let's try $-1 \le y \le 1$.

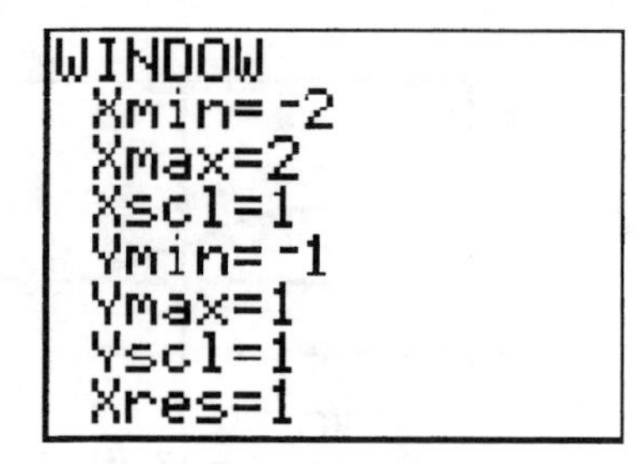

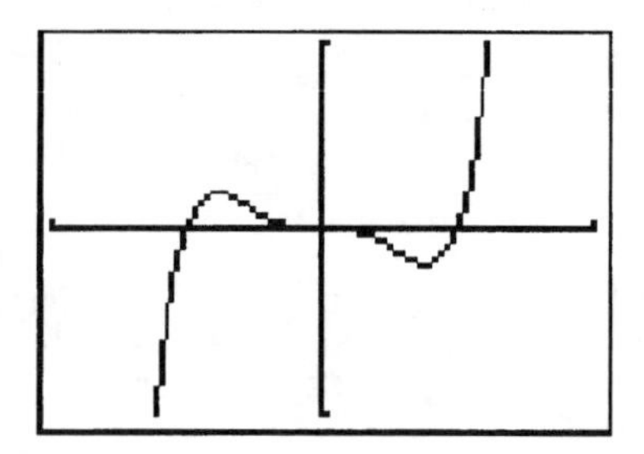

This window is much better.

Use `maximum` to find the local maximum. Note that the *x*-coordinate of the local maximum is between $x = -1$ and $x = 0$.

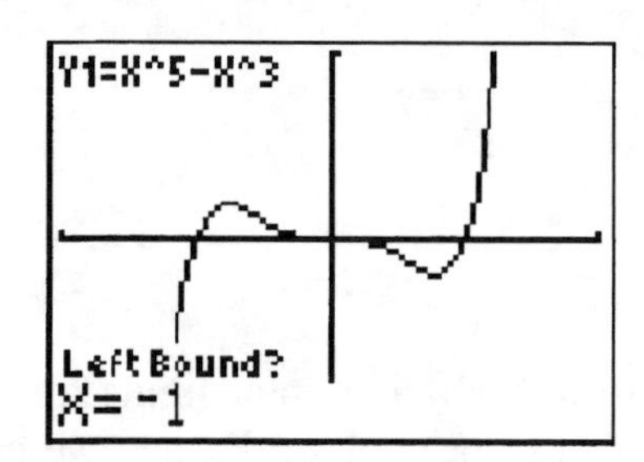

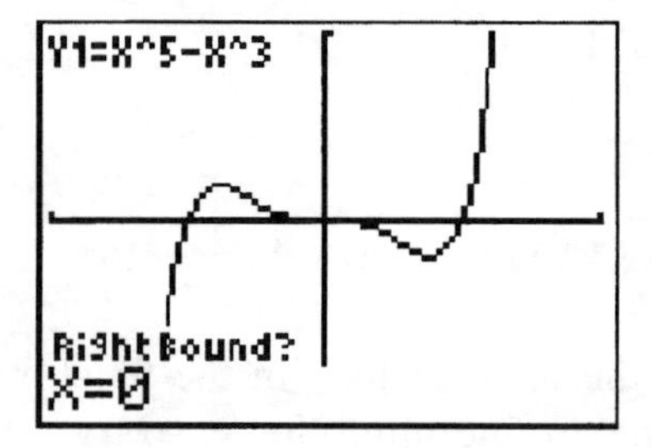

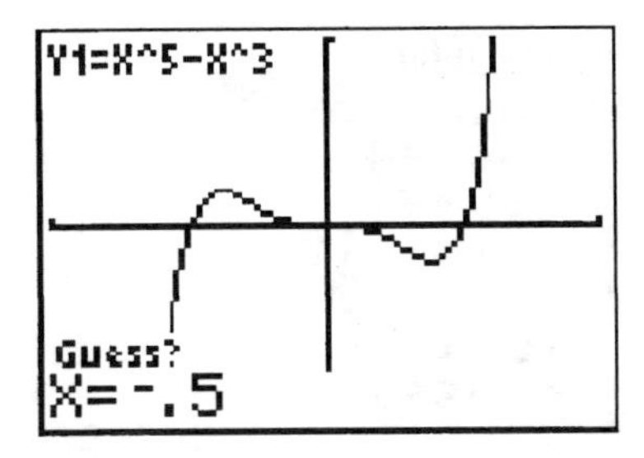

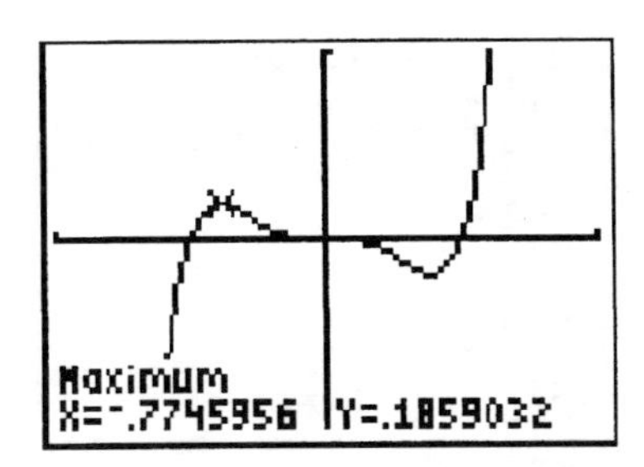

The local maximum is approximately $(-0.77, 0.19)$.

Use `minimum` to find the local minimum. Note that the x-coordinate of the local minimum is between $x = 0$ and $x = 1$.

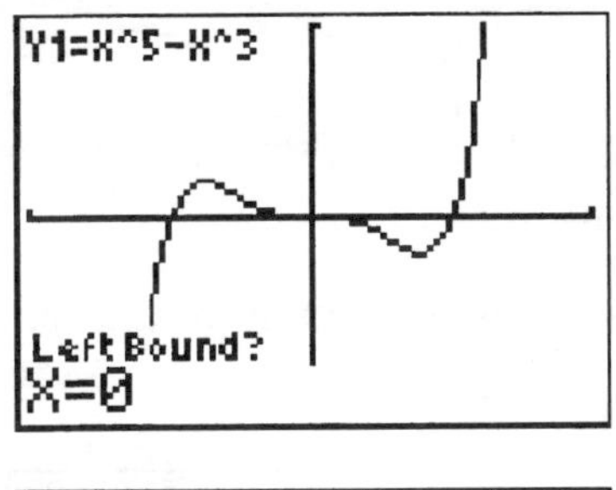

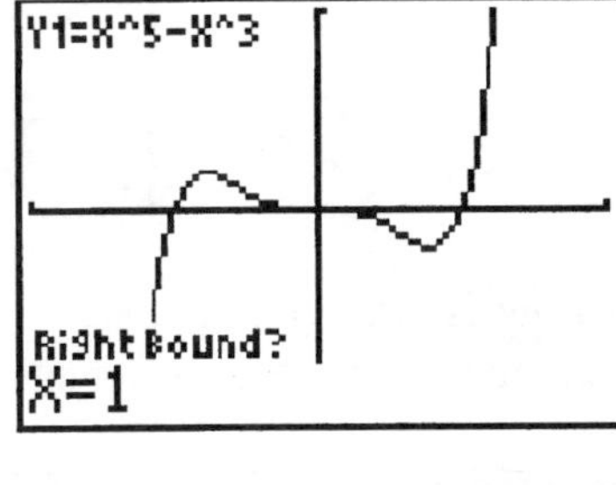

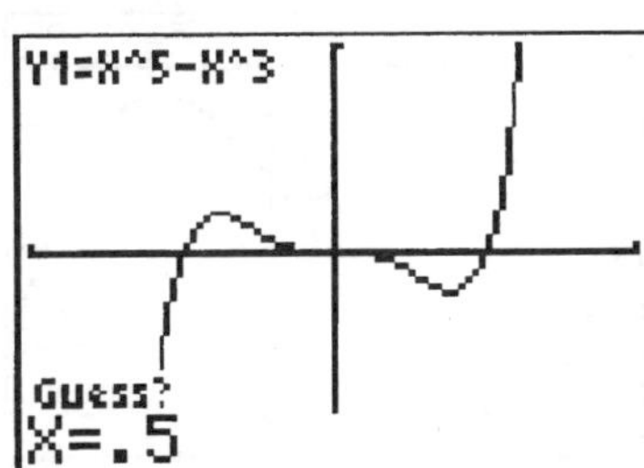

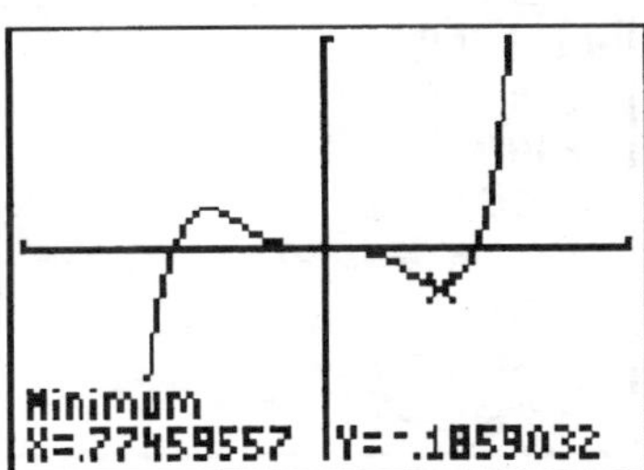

The local minimum is approximately $(0.77, -0.19)$.

Thus, f is increasing on the intervals $(-2, -0.77)$ and $(0.77, 2)$, and f is decreasing on the interval $(-0.77, 0.77)$.

73. $f(x) = -0.2x^3 - 0.6x^2 + 4x - 6 \quad (-6, 4)$

Enter the formula for f in the function editor. Go to **WINDOW** and enter limits for x. We must determine limits for y, so try $-10 \le y \le 10$. If this does not work, then we can go back to **WINDOW** and adjust the limits on y until we find a good window.

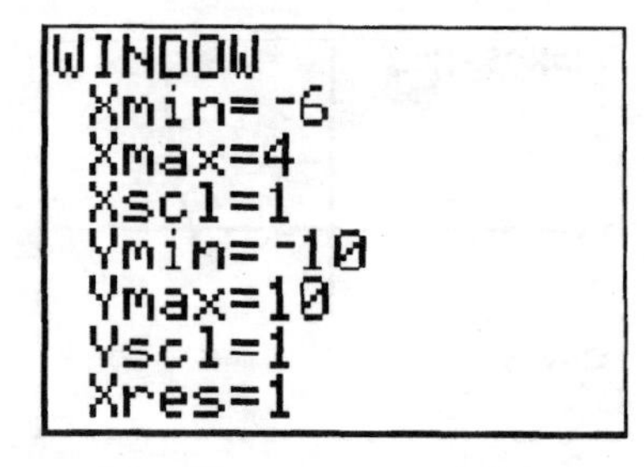

Graph the function.

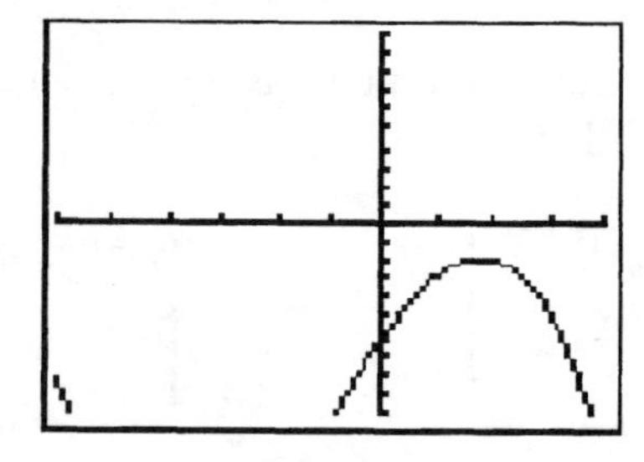

This window is not suitable, part of the graph goes below the bottom of the screen. We need to change the limits for y so let's try $-20 \le y \le 2$.

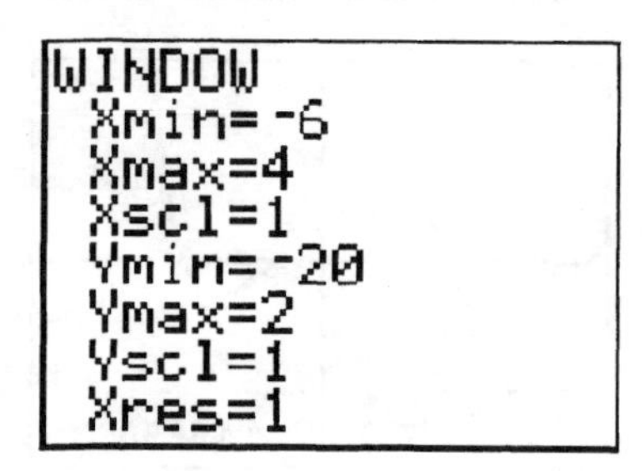

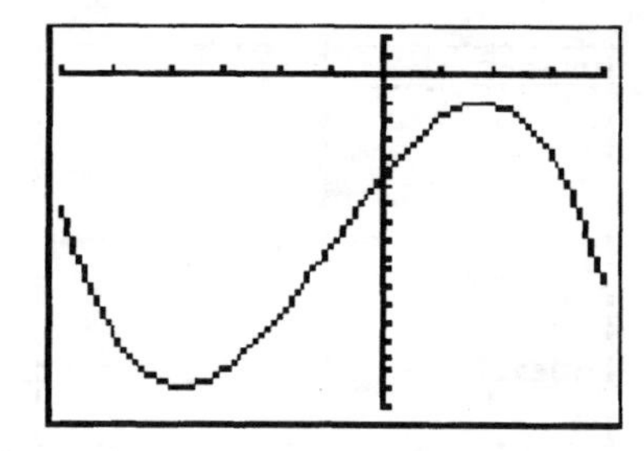

This window is much better.

Use `maximum` to find the local maximum. Note that the x-coordinate of the local maximum is between $x = 1$ and $x = 3$.

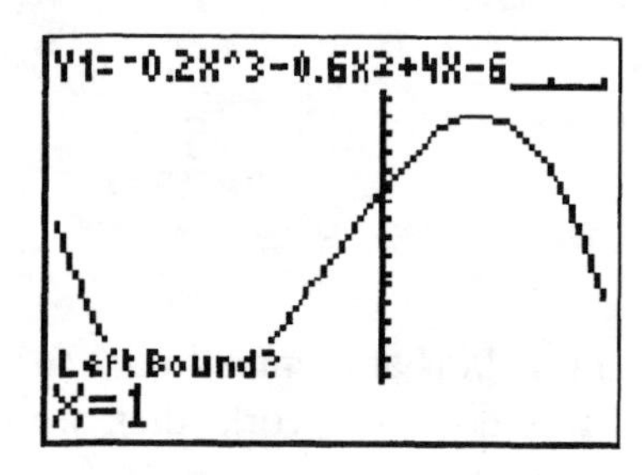

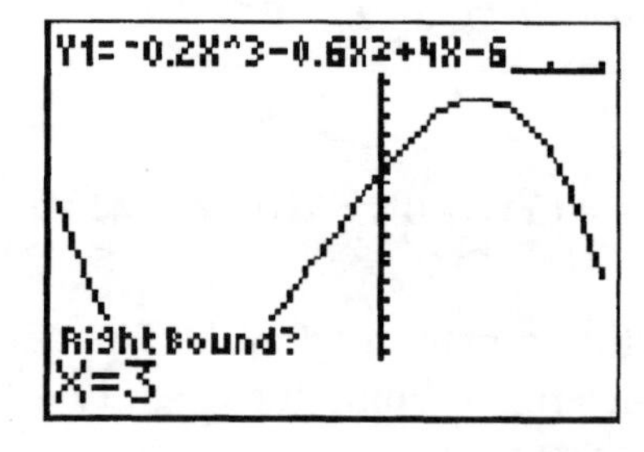

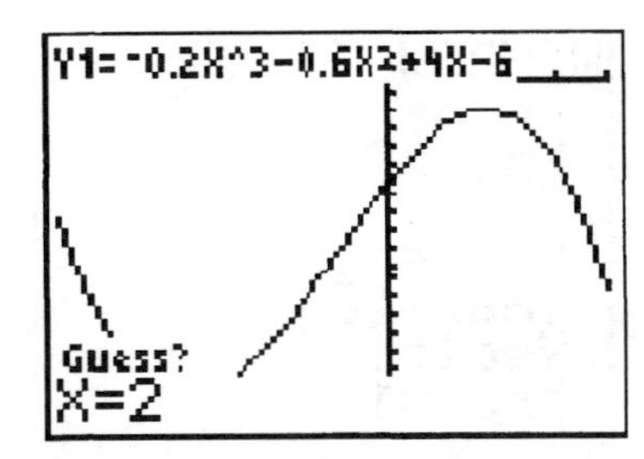

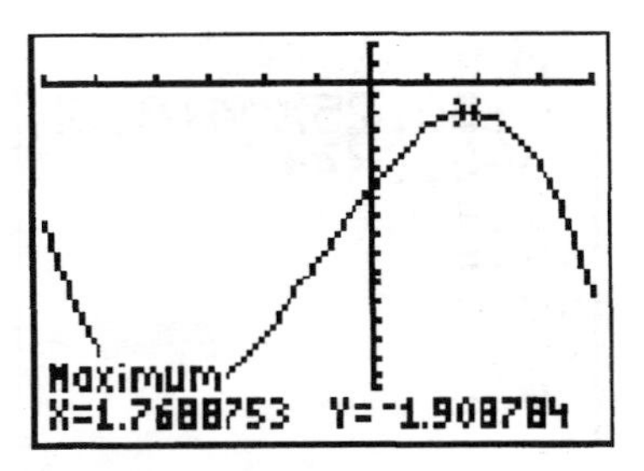

The local maximum is approximately $(1.77, -1.91)$.

Use `minimum` to find the local minimum. Note that the *x*-coordinate of the local minimum is between $x = -5$ and $x = -3$.

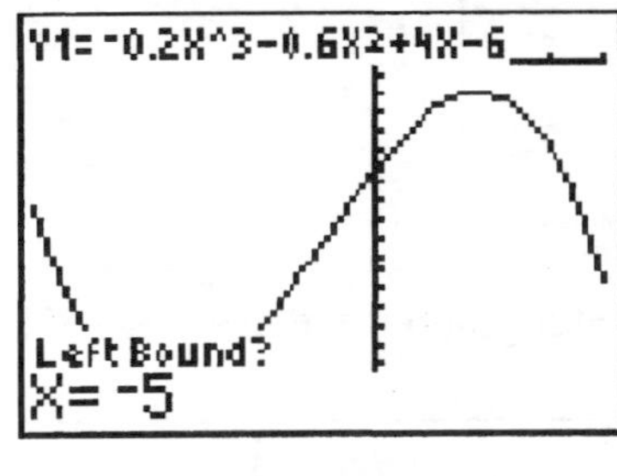

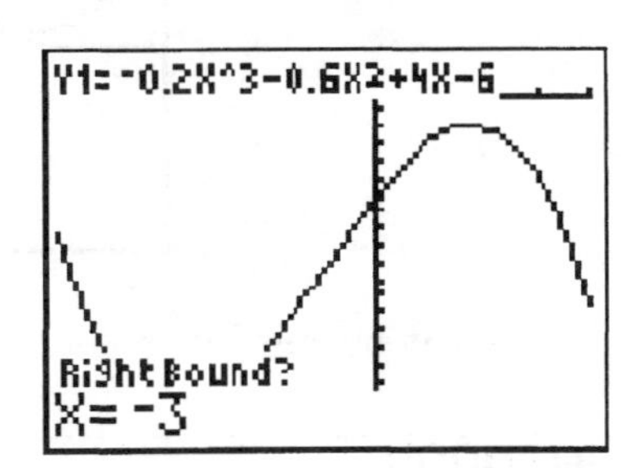

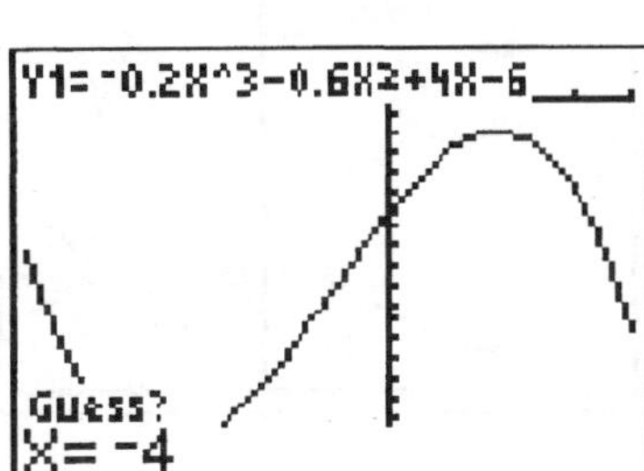

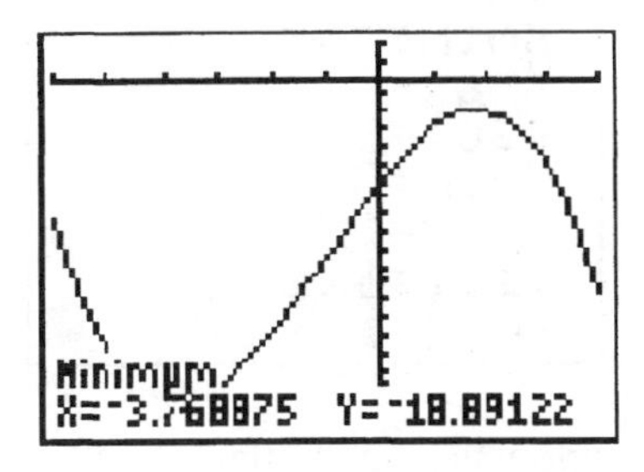

The local minimum is approximately $(-3.77, -18.89)$.

Thus, f is increasing on the interval $(-3.77, 1.77)$, and f is decreasing on the intervals $(-6, -3.77)$ and $(1.77, 4)$.

75. $f(x) = 0.25x^4 + 0.3x^3 - 0.9x^2 + 3 \quad (-3, 2)$

Enter the formula for *f* in the function editor. Go to `WINDOW` and enter limits for *x*. We must determine limits for *y*, so try $-10 \le y \le 10$. If this does not work, then we can go back to `WINDOW` and adjust the limits on *y* until we find a good window.

Plot1 Plot2 Plot3
\Y1=0.25X^4+0.3X^3-0.9X²+3
\Y2=
\Y3=
\Y4=
\Y5=
\Y6=

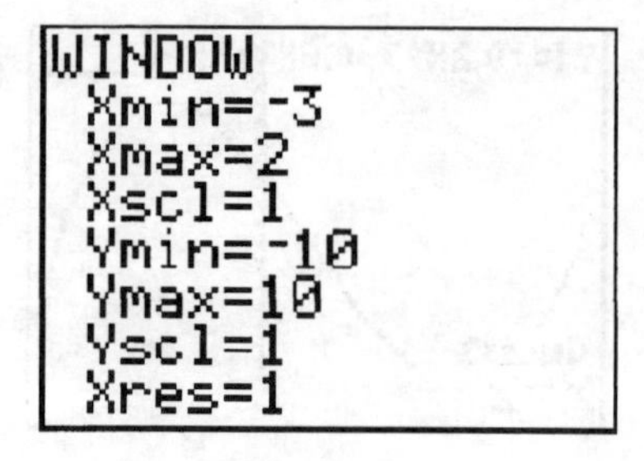

Graph the function.

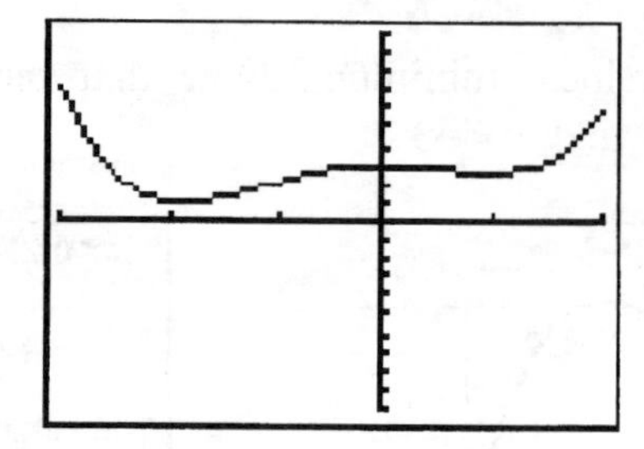

This window may be suitable, but we can obtain a better graph if we use $0 \leq y \leq 10$.

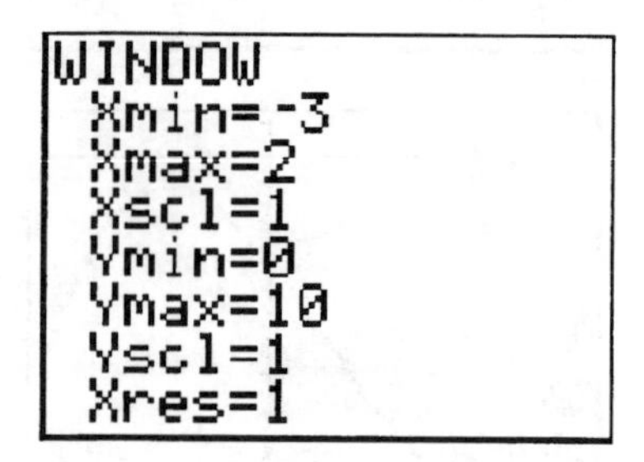

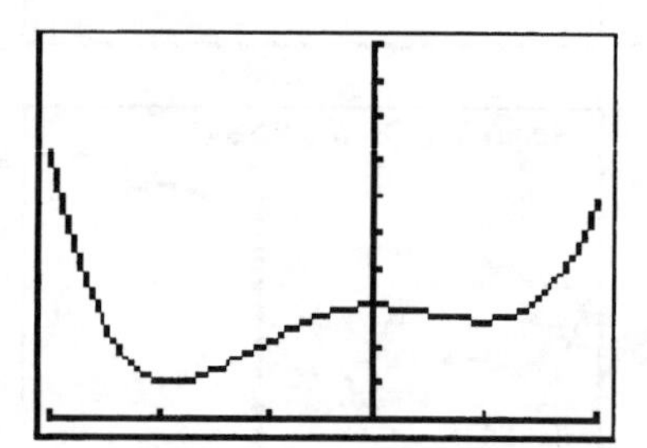

This window is much better.

Use `maximum` to find the local maximum. Note that the *x*-coordinate of the local maximum is between $x = -1$ and $x = 1$.

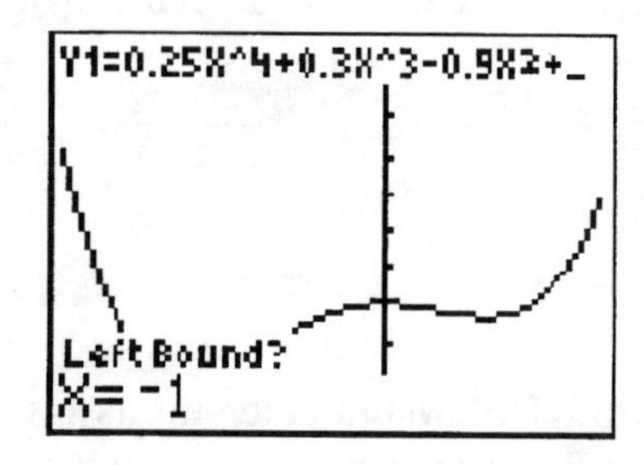

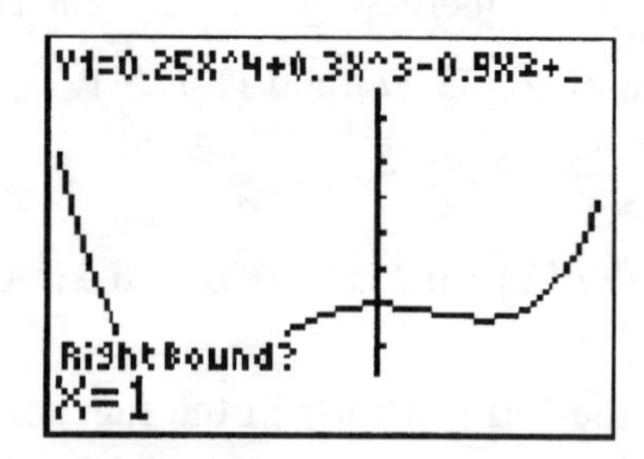

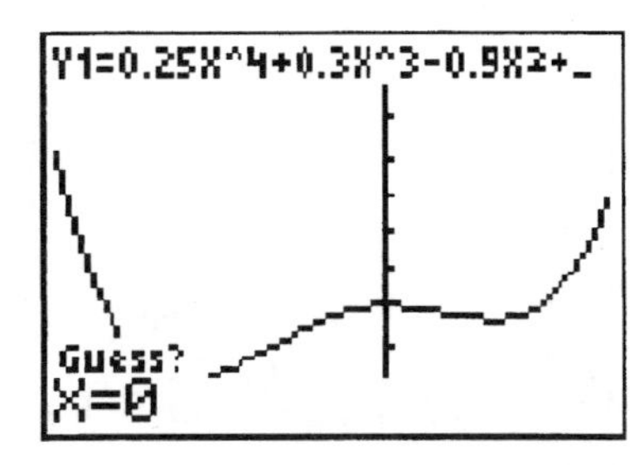

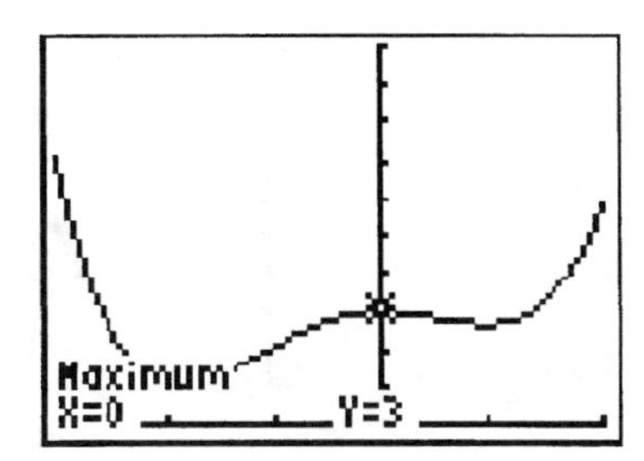

The local maximum is $(0, 3)$.

Use `minimum` to find both local minima. Note that the x-coordinate of the first local minimum is between $x = -3$ and $x = -1$.

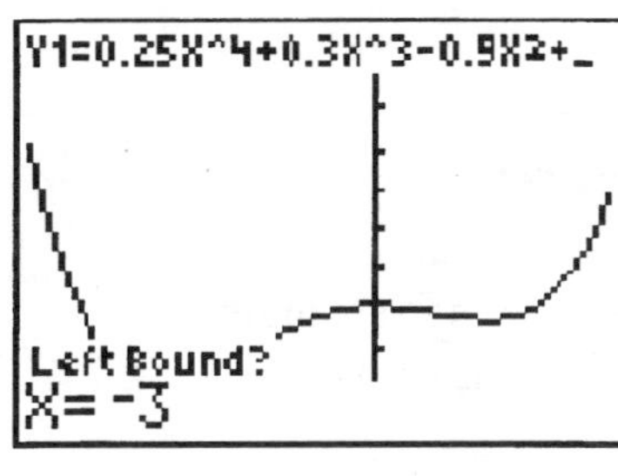

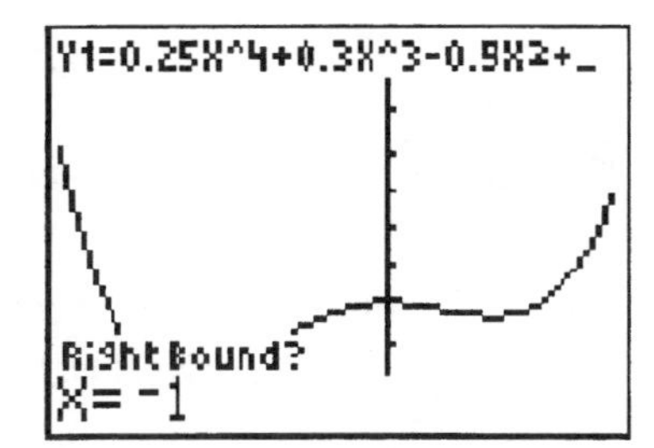

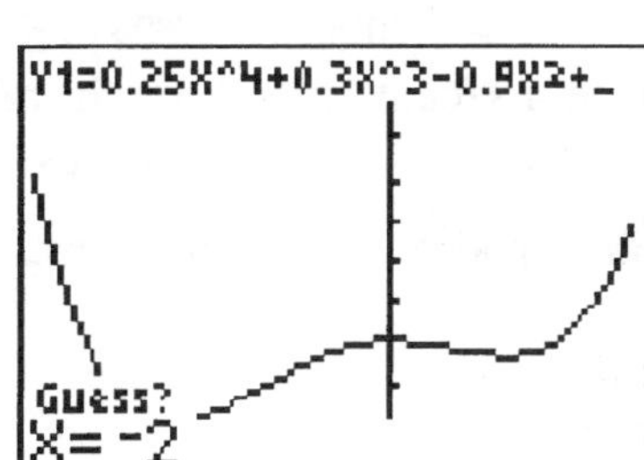

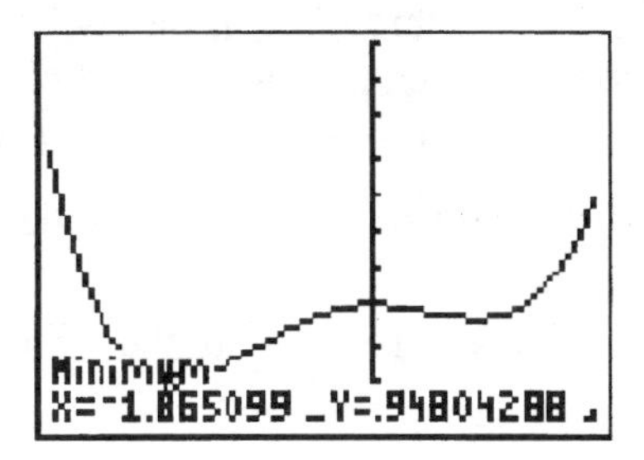

The first local minimum is approximately $(-1.87, 0.95)$.

The x-coordinate of the second local minimum is between $x = 0$ and $x = 2$.

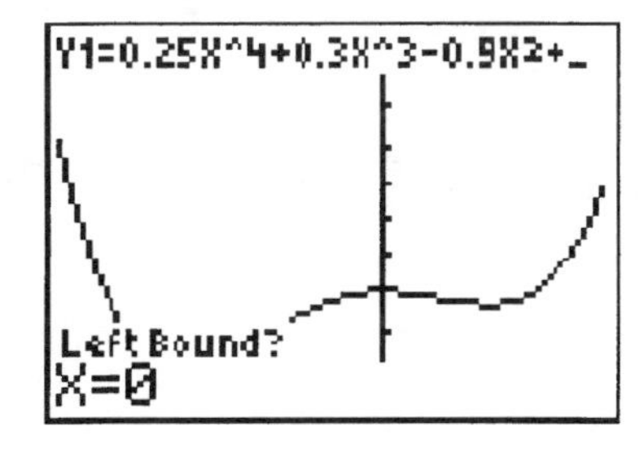

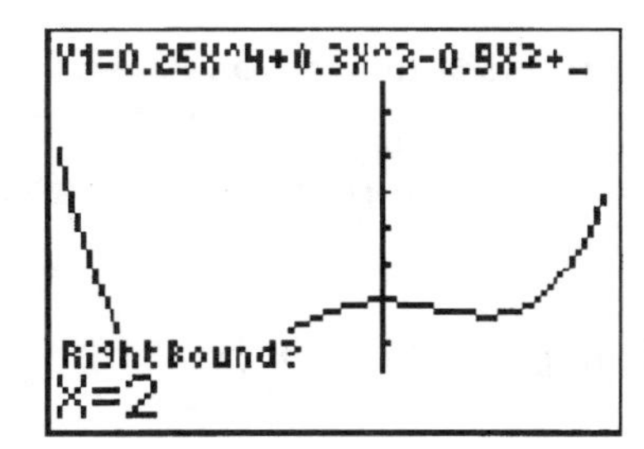

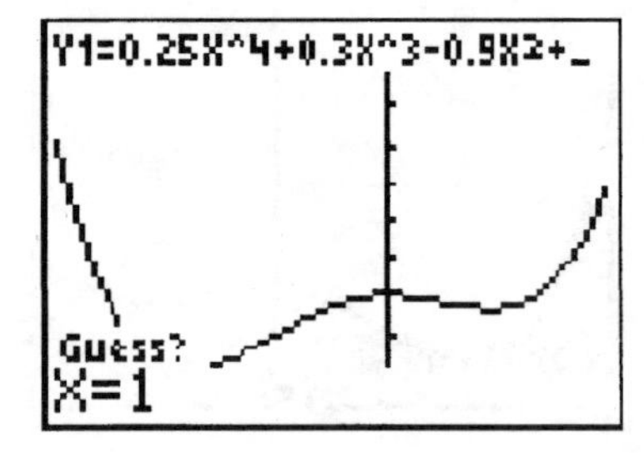

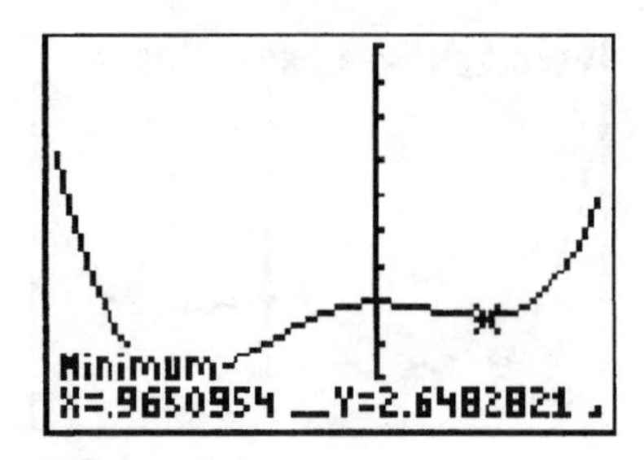

The second local minimum is approximately $(0.97, 2.65)$.

Thus, f is increasing on the intervals $(-1.87, 0)$ and $(0.97, 2)$, and f is decreasing on the intervals $(-3, -1.87)$ and $(0, 0.97)$.

77. For the function $f(x) = x^2$, compute each average rate of change:

(f) Graph each of the secant lines. Set the viewing rectangle to: $\text{Xmin} = -0.2$, $\text{Xmax} = 1.2$, $\text{Xscl} = 0.1$, $\text{Ymin} = -0.2$, $\text{Ymax} = 1.2$, and $\text{Yscl} = 0.1$.

From parts (a) through (e), the average rates of change are 1, 0.5, 0.1, 0.01, and 0.001, respectively. Each average rate of change is a slope of a secant lines through the points given in each of parts (a) through (e). In each case, one of the two points was $(0, 0)$. Using this point and each of the slopes, we can find the equations of the five secant lines.

Using $m = 1$ and $(0, 0)$ with the point-slope form of the line we obtain

$$y - 0 = 1(x - 0)$$

which simplifies to $y = x$.

Using $m = 0.5$ and $(0, 0)$ with the point-slope form of the line we obtain

$$y - 0 = 0.5(x - 0)$$

which simplifies to $y = 0.5x$.

Using $m = 0.1$ and $(0, 0)$ with the point-slope form of the line we obtain

$$y - 0 = 0.1(x - 0)$$

which simplifies to $y = 0.1x$.

Using $m = 0.01$ and $(0,\ 0)$ with the point-slope form of the line we obtain

$$y - 0 = 0.01(x - 0)$$

which simplifies to $y = 0.01x$.

Using $m = 0.001$ and $(0,\ 0)$ with the point-slope form of the line we obtain

$$y - 0 = 0.001(x - 0)$$

which simplifies to $y = 0.001x$.

Enter $f(x) = x^2$ and the equation of the secant line, $y = x$ in the function editor. Go to `WINDOW` and enter limits for *x* and *y*. Graph the two equations.

```
Plot1 Plot2 Plot3
\Y1=X²
\Y2=X
\Y3=
\Y4=
\Y5=
\Y6=
\Y7=
```

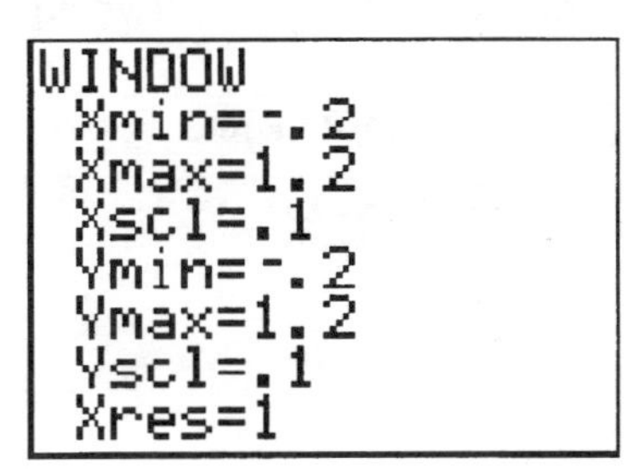

Graph $f(x) = x^2$ and the secant line $y = x$.

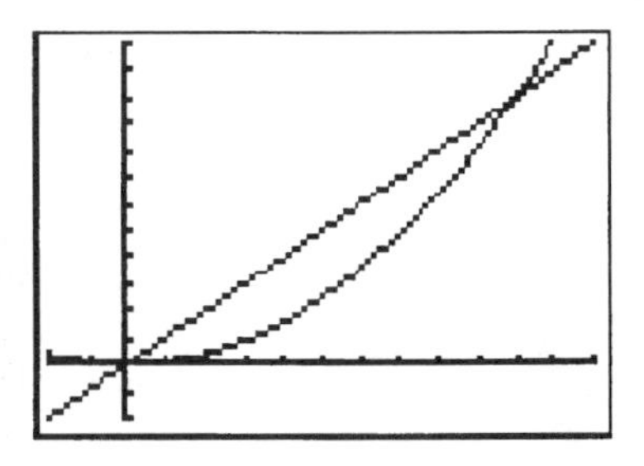

Replace the equation of the secant line $y = x$ with the next secant line, $y = 0.5x$, in the function editor. Graph the two equations.

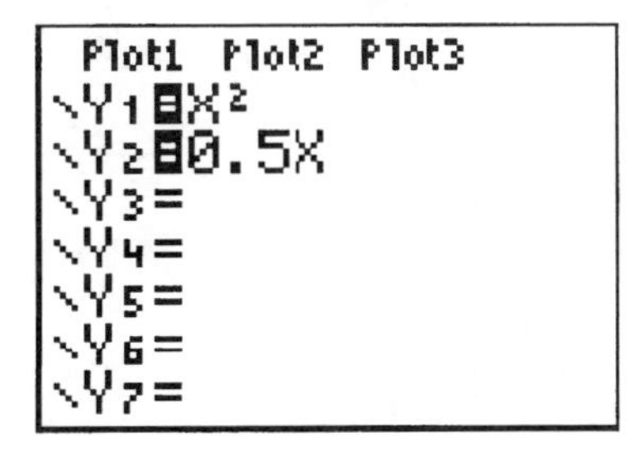

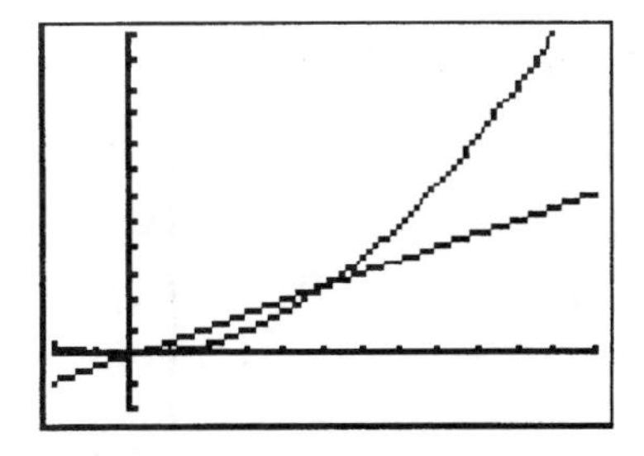

Replace the equation of the secant line $y = 0.5x$ with the next secant line, $y = 0.1x$, in the function editor. Graph the two equations.

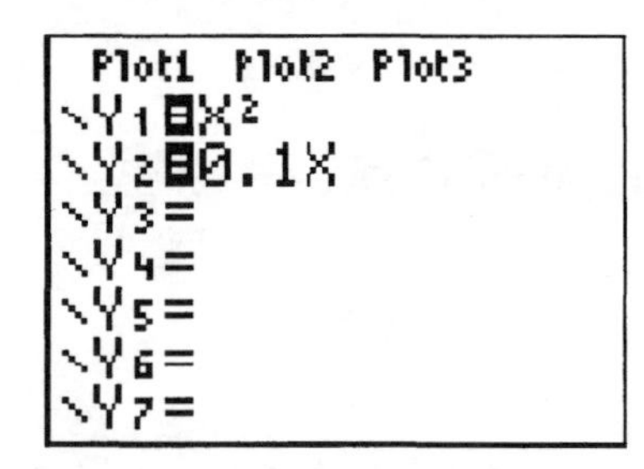

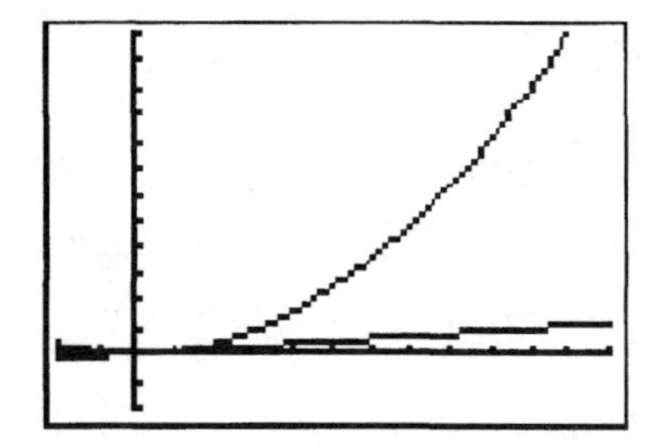

Replace the equation of the secant line $y = 0.1x$ with the next secant line, $y = 0.01x$, in the function editor. Graph the two equations.

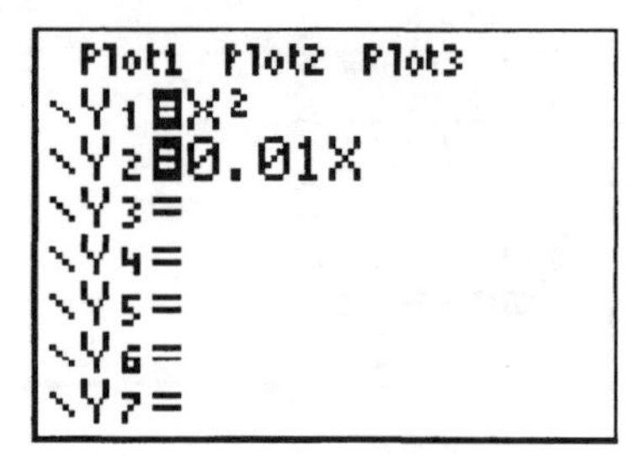

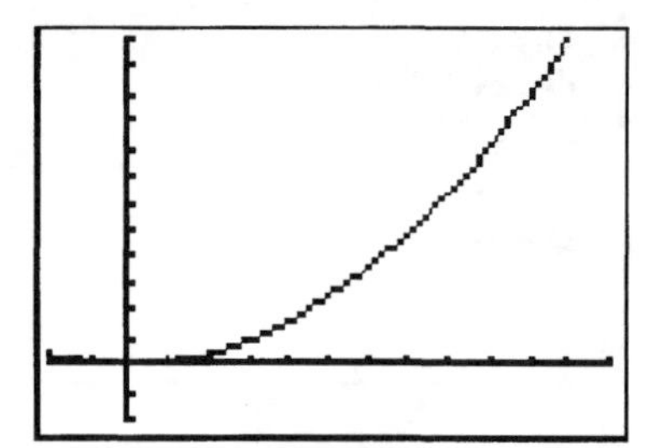

It looks as though the calculator only graphed $f(x) = x^2$. The calculator did graph the line, but we cannot distinguish the line from the x-axis. To see the line, you must turn the axes "off."

[2nd] [ZOOM] [▾] [▾] [▾] [▸] [ENTER]

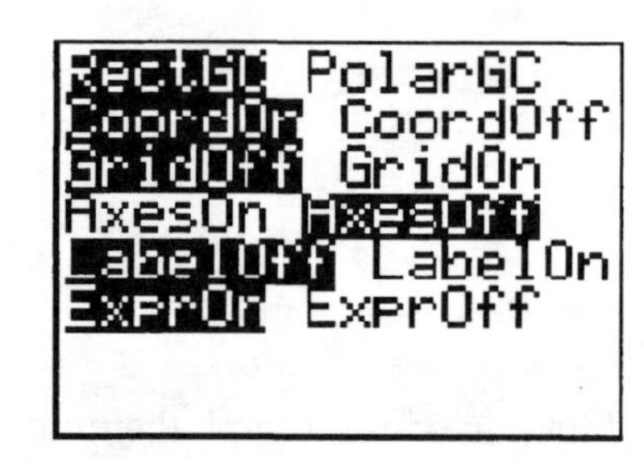

Regraph the two equations.

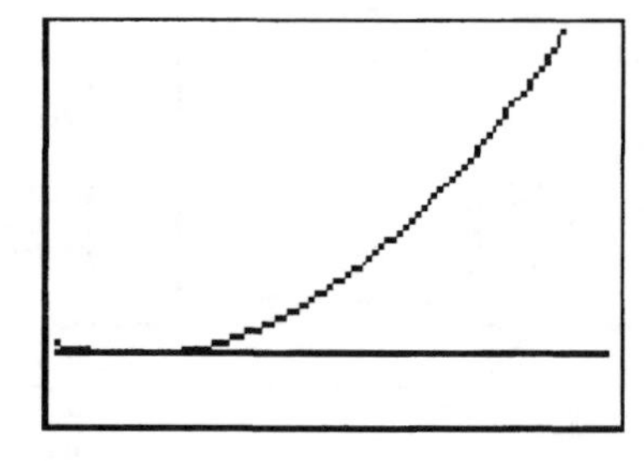

Replace the equation of the secant line $y = 0.01x$ with the secant line $y = 0.001x$ in the function editor. Graph the two equations.

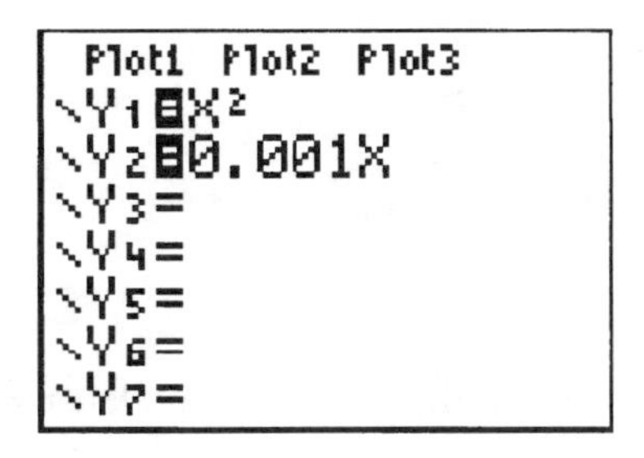

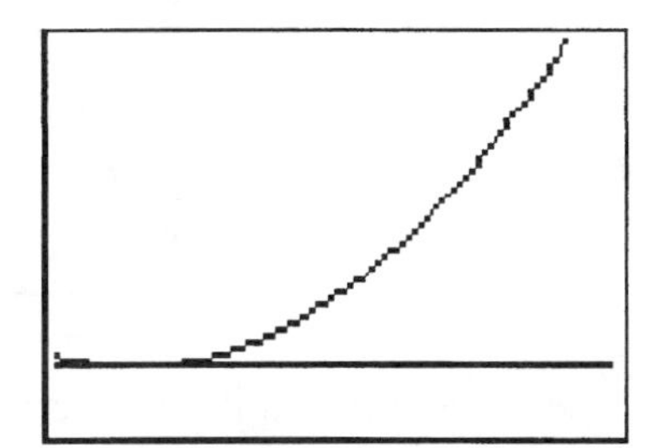

Turn the axes "on" when you are done.

79. **Motion of a Golf Ball** A golf ball is hit with an initial velocity of 130 feet per second at an inclination of 45° to the horizontal. In physics, it is established that the height h of the golf ball is given by the function

$$h(x) = \frac{-32x^2}{130^2} + x$$

where x is the horizontal distance that the golf ball has traveled.

(e) Graph the function $h = h(x)$.

Enter the formula for h in the function editor. Go to `WINDOW` and enter limits for x and y. We are not given any limits for either variable. For this problem to make sense, we must have $x \geq 0$. Let's try the following window.

```
Plot1 Plot2 Plot3
\Y1=(-32X²)/130²
+X
\Y2=
\Y3=
\Y4=
\Y5=
\Y6=
```

```
WINDOW
 Xmin=0
 Xmax=600
 Xscl=50
 Ymin=0
 Ymax=150
 Yscl=25
 Xres=1
```

Graph the function.

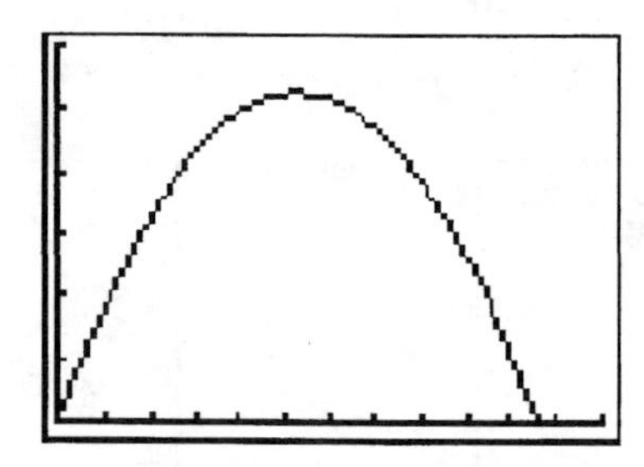

(f) Use a graphing utility to determine the distance that the ball has traveled when the height of the ball is 90 feet.

One way to solve this problem is to graph the horizontal line $g(x)=90$ on the same graph as $h(x)$, and use `intersect` to find the point(s) where the two functions intersect. The `intersect` function requires three inputs: the equation corresponding to the first curve; the equation corresponding to the second curve; and a guess for the x-coordinate of the point of intersection. The `intersect` function can be found under the `[CALC]` menu.

Enter the function $g(x)=90$ in the function editor and graph both functions.

```
Plot1 Plot2 Plot3
\Y1=(-32X²)/130²
+X
\Y2=90
\Y3=
\Y4=
\Y5=
\Y6=
```

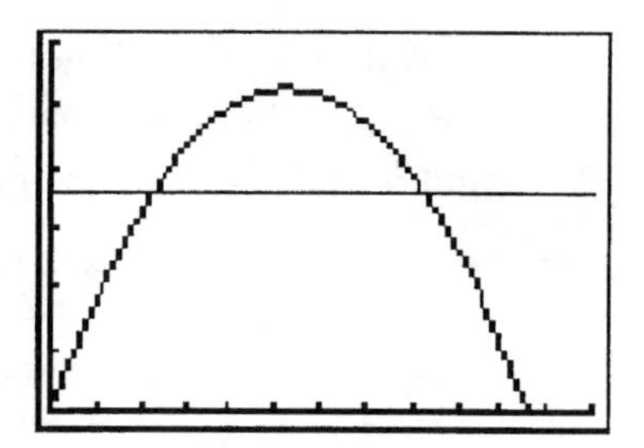

There are two points where the two functions intersect. Notice that the x-coordinate of the first (or left) point is close to $x=100$, while the x-coordinate of the second (or right) point is close to $x=450$.

Find the first point of intersection.

[2nd] [TRACE]

Select intersect.

[5]

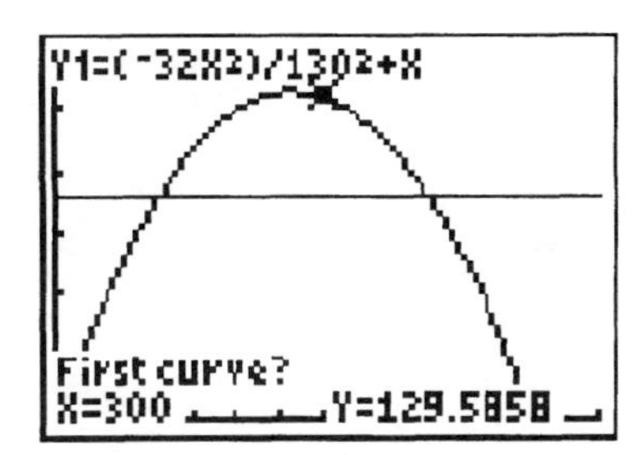

Since the cursor is on first curve (the parabola) we can just press [ENTER] to select that curve.

[ENTER]

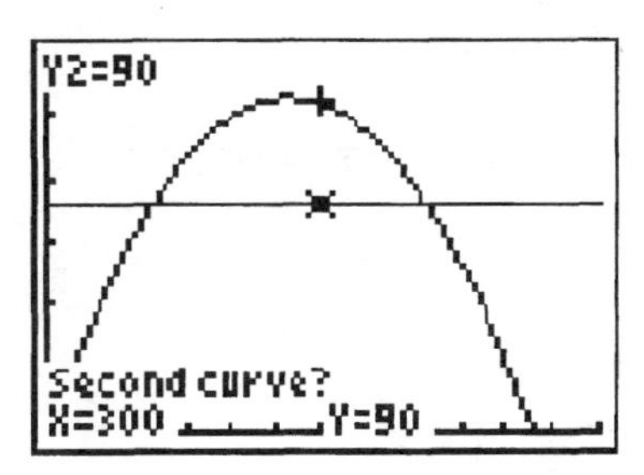

The cursor is now on the second curve (the line), so press [ENTER] to select that curve.

[ENTER]

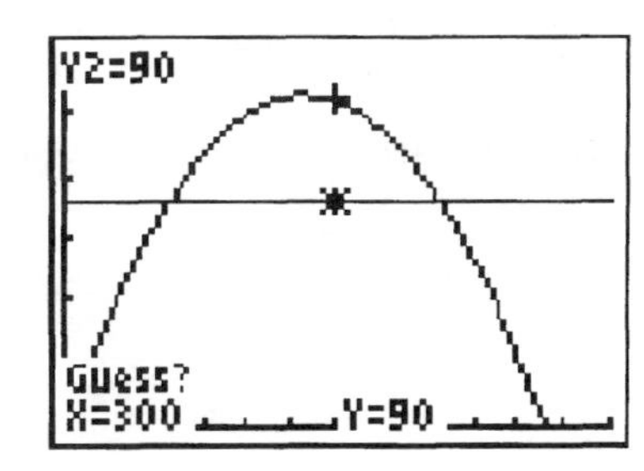

Now, input a guess for the x-coordinate of the first intersection point. Let's use $x = 100$.

[1] [0] [0]

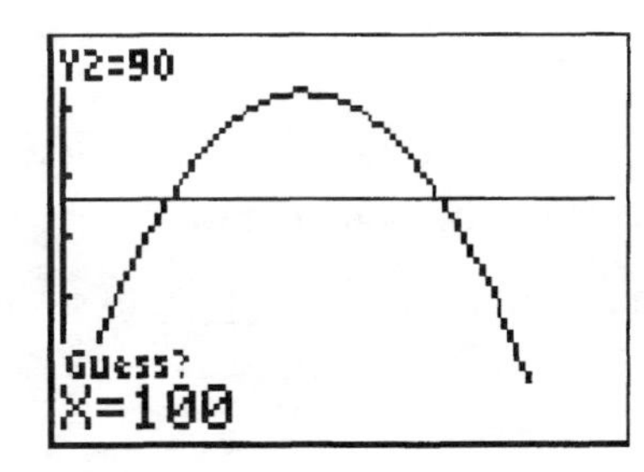

Press [ENTER] to find the first intersection point.

[ENTER]

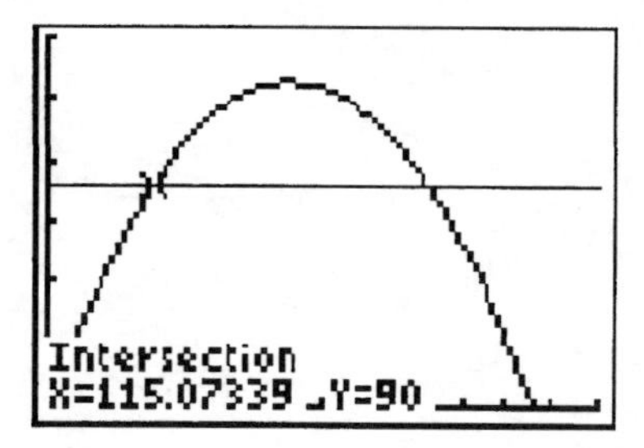

The coordinates of the point are approximately $(115.07, 90)$, thus the golf ball has traveled a horizontal distance of approximately 115.07 feet when it first reaches a height of 90 feet.

Now, find the second point of intersection.

[2nd] [TRACE]

CALCULATE
1:value
2:zero
3:minimum
4:maximum
5:intersect
6:dy/dx
7:∫f(x)dx

Select `intersect`.

[5]

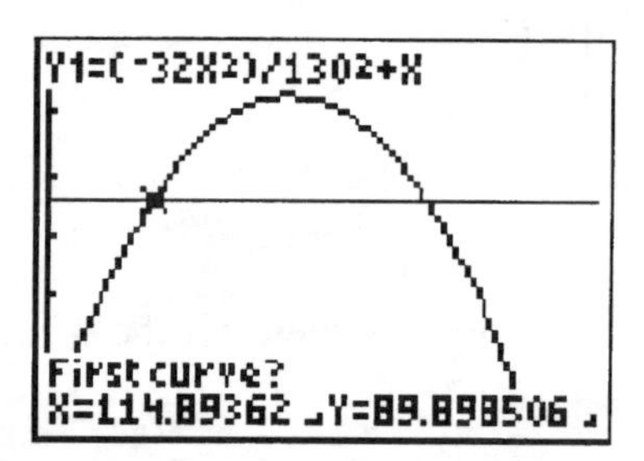

Since the cursor is on the first intersection point, it is hard to tell which curve it is on. Notice the equation of $h(x)$ at the top of the screen, this tells us the cursor is on the parabola, so press [ENTER] to select $h(x)$.

[ENTER]

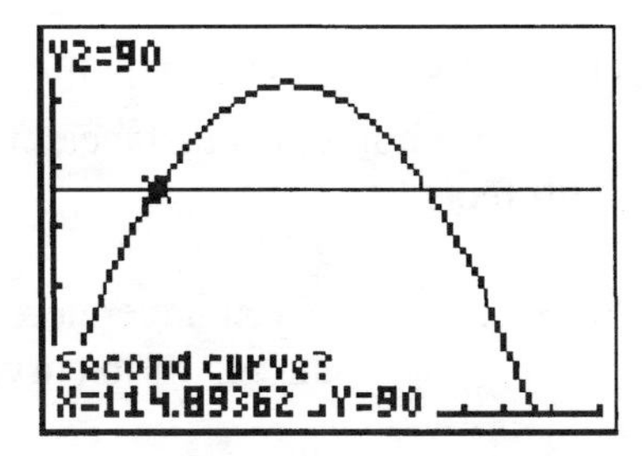

Again, the cursor is on the first intersection point, but we know that it is on the horizontal line by looking at the equation listed at the top of the screen. Press [ENTER] to select that curve.

[ENTER]

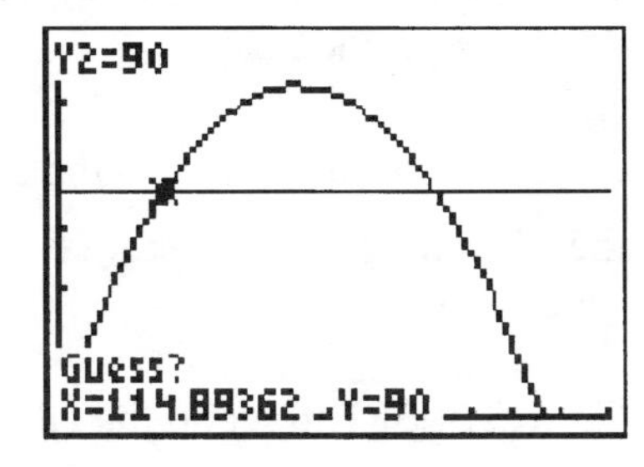

Now, input a guess for the x-coordinate of the second intersection point. Let's use $x = 450$.

[4] [5] [0]

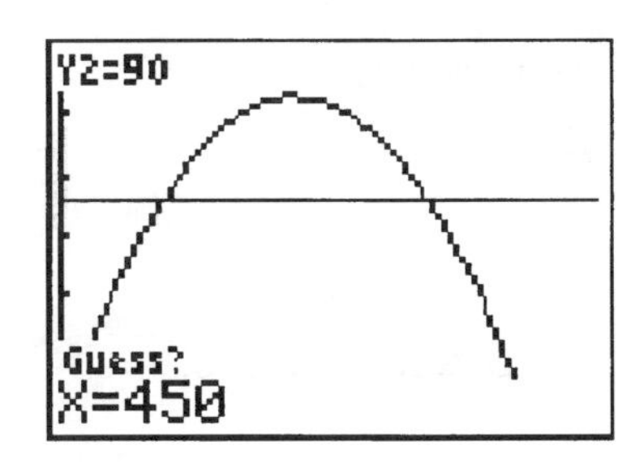

Press [ENTER] to find the second intersection point.

[ENTER]

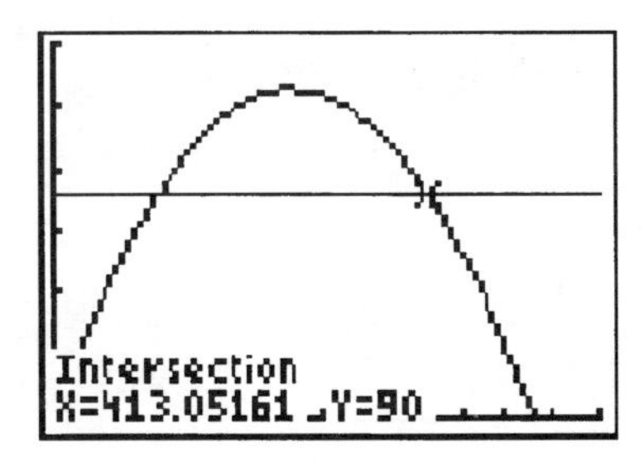

The coordinates of the point are approximately $(413.05, 90)$, thus the golf ball has traveled a horizontal distance of approximately 413.05 feet when it next reaches a height of 90 feet.

NOTE: If you have more than two curves graphed, use the [▲] or [▼] keys to move the cursor to curve you wish to mark.

(g) Create a TABLE with TblStart = 0 and ΔTbl = 25.

This problem is asking us to generate a table of values for our function $h(x)$. The values for x start with $x = 0$, and the rest of the values of x are obtained by adding the increment 25 to the previous x.

First, be sure that $h(x)$ is the only function in the function editor (so if $g(x) = 90$ is still there from part (f), remove it), or, alternatively, deselect all other functions in the function editor.

Go to TBLSET to set up the table.

[2nd] [WINDOW]

```
TABLE SETUP
 TblStart=0
 ΔTbl=1
Indpnt: Auto Ask
Depend: Auto Ask
```

Set TblStart to zero.

[0] [ENTER]

```
TABLE SETUP
 TblStart=0
 ΔTbl=1
Indpnt: Auto Ask
Depend: Auto Ask
```

Set ΔTbl to 25.

[2] [5] [ENTER]

```
TABLE SETUP
 TblStart=0
 ΔTbl=25
Indpnt: Auto Ask
Depend: Auto Ask
```

Set Indpnt to Auto.

[ENTER]

```
TABLE SETUP
 TblStart=0
 ΔTbl=25
Indpnt: Auto Ask
Depend: Auto Ask
```

Generate the table of values.

[2nd] [GRAPH]

X	Y1
0	0
25	23.817
50	45.266
75	64.349
100	81.065
125	95.414
150	107.4

X=0

To scroll through the table, use the [▼] or [▲] keys.

[▼] [▼] [▼] [▼] [▼] [▼] [▼] [▼] [▼] [▼] [▼] [▼] [▼]

X	Y1
175	117.01
200	124.26
225	129.14
250	131.66
275	131.8
300	129.59
325	125

X=325

(h) To the nearest 25 feet, how far does the ball travel before it reaches a maximum height? What is the maximum height?

Continue to scroll through the table generated in part (g).

[▼] [▼] [▼] [▼] [▼] [▼] [▼]

X	Y1
350	118.05
375	108.73
400	97.041
425	82.988
450	66.568
475	47.781
500	26.627

X=500

[▼] [▼]

X	Y1
400	97.041
425	82.988
450	66.568
475	47.781
500	26.627
525	3.1065
550	-22.78

X=550

Looking at the table we see that the largest value for h is approximately 132, which occurs when x is 275. In other words, the golf ball reaches a maximum height of approximately 132 feet after it has traveled a horizontal distance of approximately 275 feet.

(i) Adjust the value of ΔTbl until you determine the distance, to within 1 foot, that the ball travels before it reaches a maximum height.

Return to TBLSET and set ΔTbl to 1 and set TblStart to 250.

Return to the table and find the maximum.

X	Y1
261	132.01
262	132.02
263	132.03
264	132.03
265	132.03
266	132.02
267	132.01

X=267

Notice that the height h appears to be the same when x is 263, 264, and 265. In order to see a more accurate value for h, move the cursor over to the second column and scroll through the values for h until you find the largest value.

X	Y1
261	132.01
262	132.02
263	132.03
264	132.03
265	132.03
266	132.02
267	132.01

Y1=132.031242604

Looking at the table we see that the largest value for h is 132.031, which occurs when x is 264. In other words, the golf ball reaches a maximum height of approximately 132.031 feet after it has traveled a horizontal distance of approximately 264 feet.

81. **Constructing an Open Box** An open box with a square base is to be made from a square piece of cardboard 24 inches on a side by cutting our a square from each corner and turning up the sides (see the figure on Page 135).

(d) Graph $V = V(x)$. For what value of x is V largest?

The function for V is given by $V(x) = x(24-2x)^2$. Enter the formula for V in the function editor. Go to `WINDOW` and enter limits for x and y. We are not given any limits for either variable, although for this problem to make sense, we must have $0 \le x \le 12$. Let's try the following window.

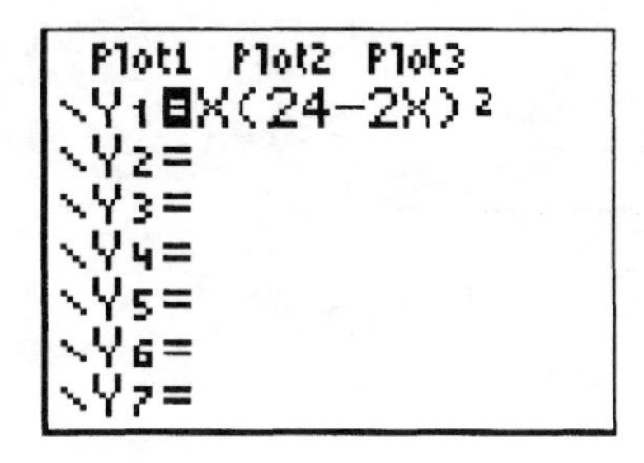

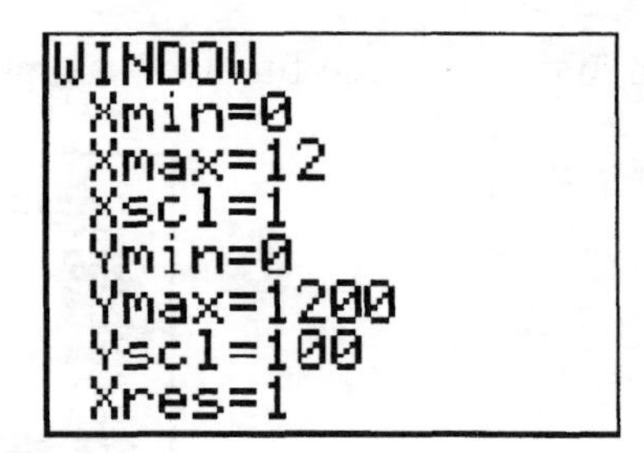

Graph $V(x)$.

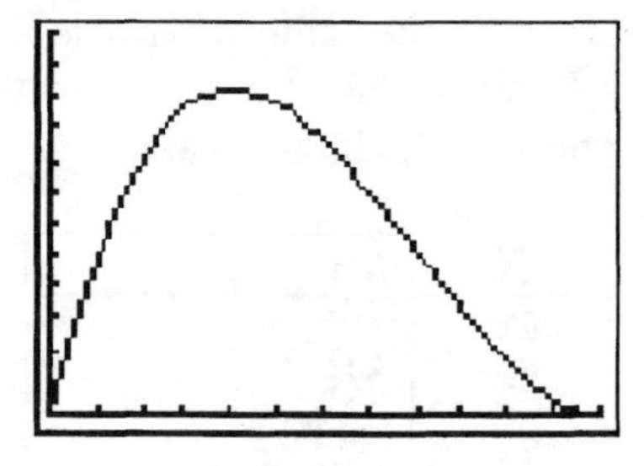

Find the local maximum. Notice that the maximum is between $x = 3$ and $x = 6$.

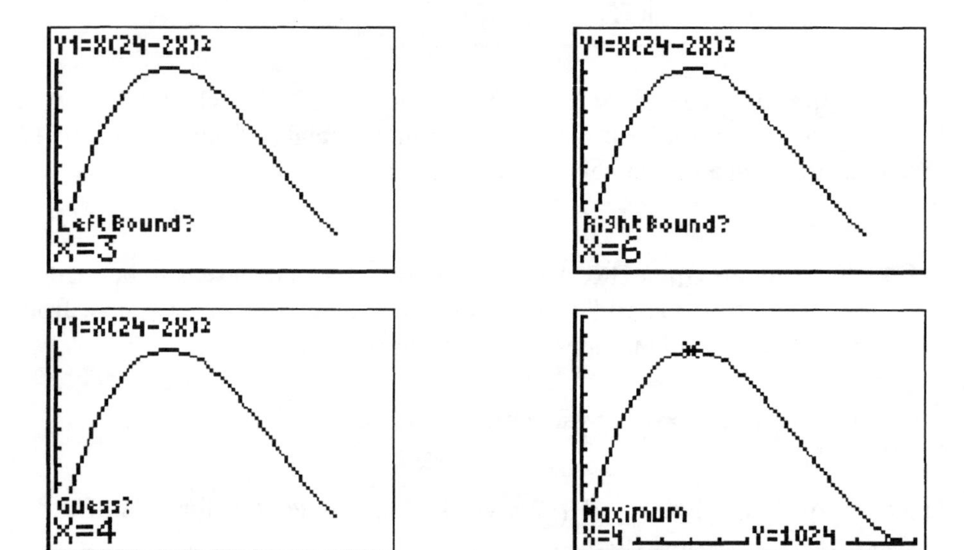

The local maximum is $(4, 1024)$. Thus, for a maximum volume of 1,024 square inches, cut out squares whose sides are 4 inches.

83. **Minimum Average Cost** The average cost of producing x riding lawn mowers per hour is given by

$$\overline{C}(x) = 0.3x^2 + 21x - 251 + \frac{2500}{x}$$

(a) Use a graphing utility to graph $\overline{C}$.

Enter the formula for $\overline{C}$ in the function editor. Go to WINDOW and enter limits for x and y. We are not given any limits for either variable, although for this problem to make sense, we must have $x > 0$. Let's try the following window.

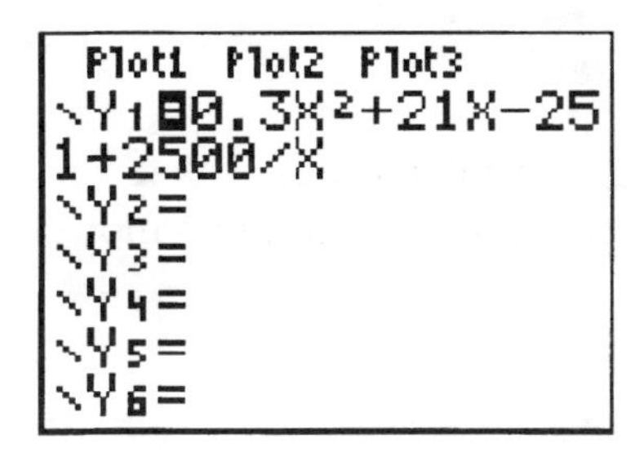

WINDOW
Xmin=0
Xmax=100
Xscl=10
Ymin=0
Ymax=5000
Yscl=1000
Xres=1

Graph $\overline{C}$.

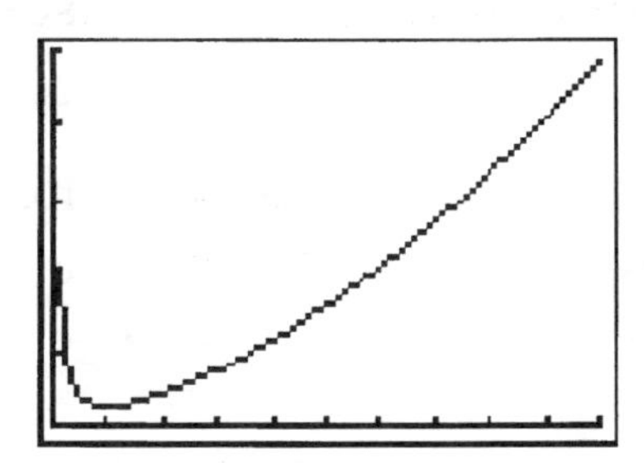

(b) Determine the number of riding lawn mowers to produce in order to minimize the average cost.

Find the local minimum. Notice that the minimum is between $x = 5$ and $x = 15$.

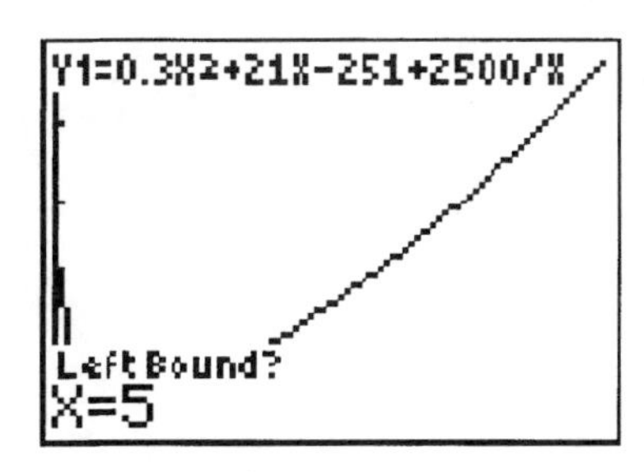

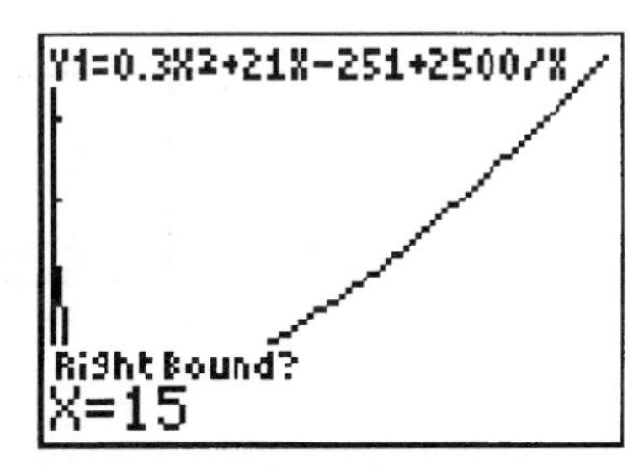

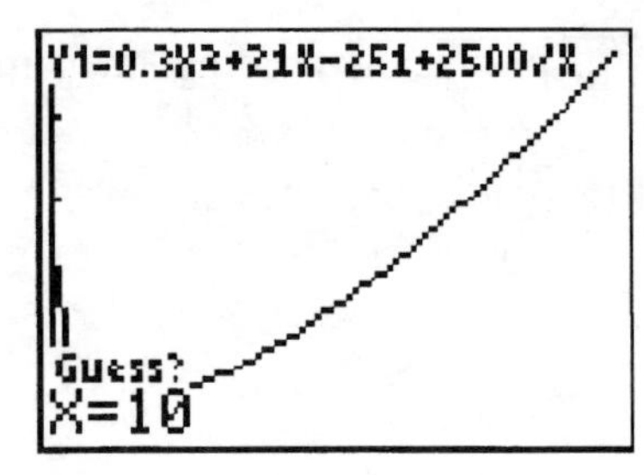

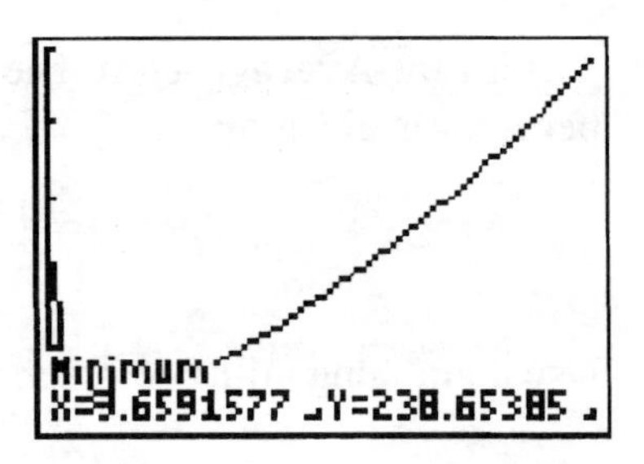

The local minimum is approximately $(9.66, 238.65)$. Since they cannot produce 9.66 riding mowers, they should produce either 9 or 10 riding mowers.

(c) What is the minimum average cost?

If they produce 9 riding mowers, the average cost is \$240.08, but if they produce 10 riding mowers the average cost is \$239.00. They should produce 10 riding mowers for an average cost of \$239.00.

Section 10.4 Library of Functions; Piecewise-Defined Functions

45. **Exploration** Graph $y = x^2$. Then on the same screen graph $y = x^2 + 2$, followed by $y = x^2 + 4$, followed by $y = x^2 - 2$. What pattern do you observe? Can you predict the graph of $y = x^2 - 4$? Of $y = x^2 + 58$.

Enter $y = x^2$ in the function editor. Go to `WINDOW` and enter limits for *x* and *y*. Let's use the standard window $-10 \le x \le 10$ and $-10 \le y \le 10$.

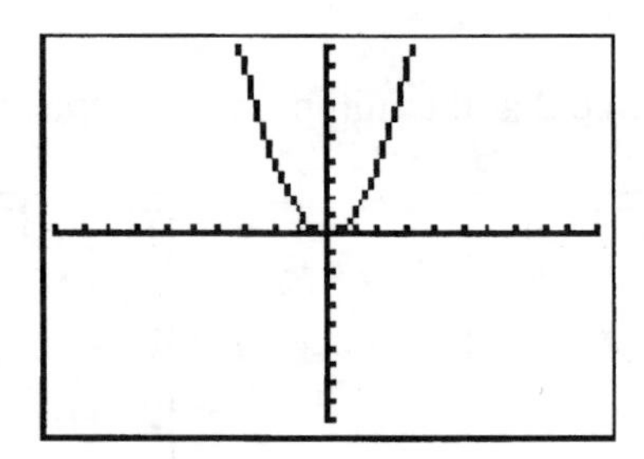

Return to the function editor and enter $y = x^2 + 2$ into y2 and graph the two equations.

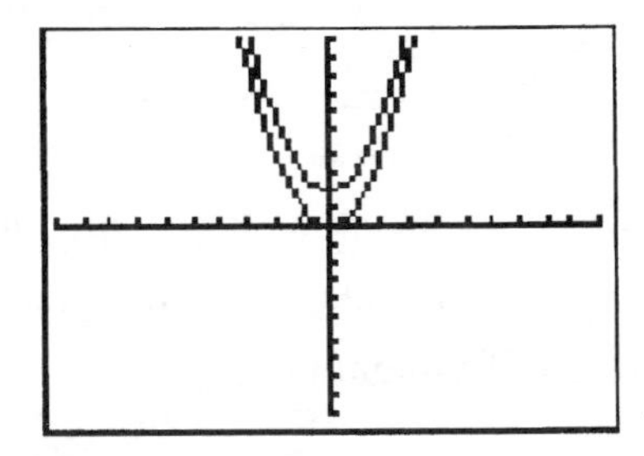

Notice that the new graph is another parabola. The only difference is its position on the y-axis. The graph of $y = x^2 + 2$ is obtained by shifting the graph of $y = x^2$ up two units.

Return to the function editor and enter $y = x^2 + 4$ into y2 and graph the two equations.

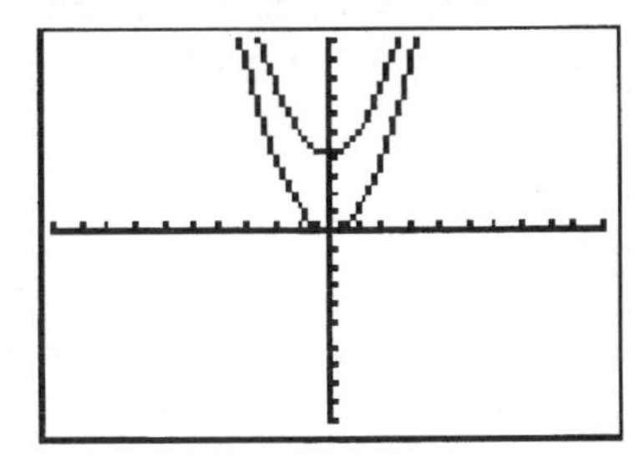

Notice that the new graph is another parabola. The only difference is its position on the y-axis. The graph of $y = x^2 + 4$ is obtained by shifting the graph of $y = x^2$ up four units.

Return to the function editor and enter $y = x^2 - 2$ into y2 and graph the two equations.

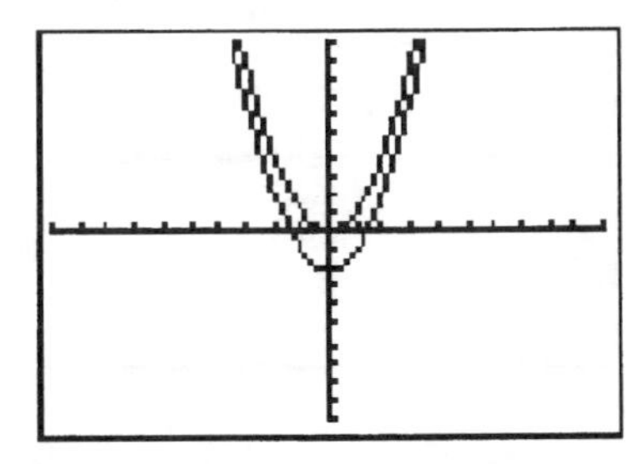

Notice that the new graph is another parabola. Again, the only difference is its position on the y-axis. The graph of $y = x^2 - 2$ is obtained by shifting the graph of $y = x^2$ down two units.

It appears that the graph of $y = x^2 + k$, $k > 0$, is obtained by shifting the graph of $y = x^2$ up k units, and that the graph of $y = x^2 - k$, $k > 0$, is obtained by shifting the graph of $y = x^2$ down k units

Based on these observations, we predict that the graph of $y = x^2 - 4$ is obtained by shifting the graph of $y = x^2$ down four units, while the graph of $y = x^2 + 58$ is obtained by shifting the graph of $y = x^2$ up fifty-eight units.

47. **Exploration** Graph $y = x^2$. Then on the same screen graph $y = -x^2$. What pattern do you observe? Now try $y = |x|$ and $y = -|x|$. What do you conclude?

Enter $y = x^2$ in the function editor. Go to `WINDOW` and enter limits for x and y. Let's use the standard window $-10 \leq x \leq 10$ and $-10 \leq y \leq 10$.

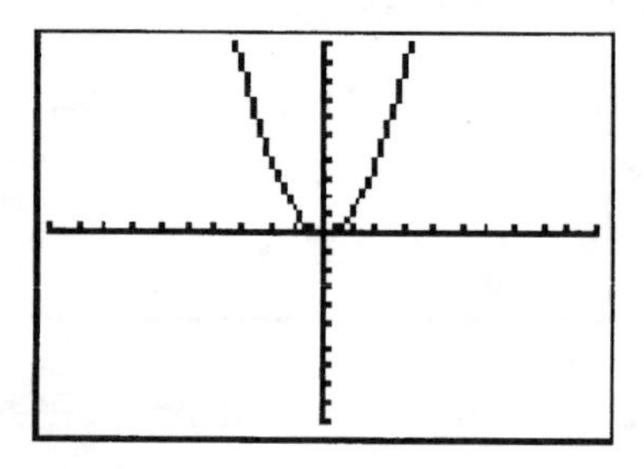

Return to the function editor and enter $y = -x^2$ into `y2` and graph the two equations.

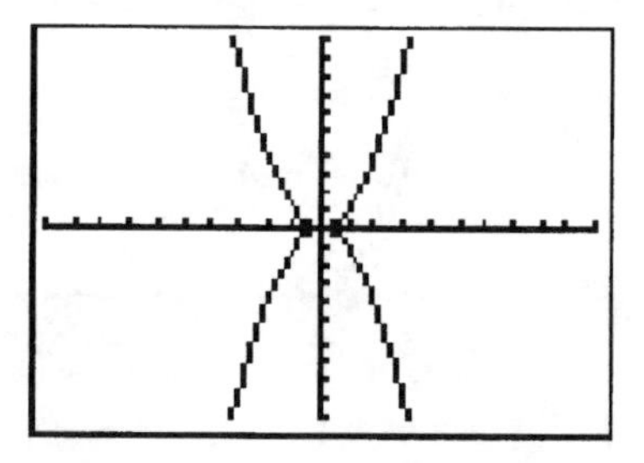

Notice that the new graph is another parabola. The only difference is the parabola is on the opposite side of the x-axis, and it opens down. In other words, the graph of $y = -x^2$ is obtained by reflecting the graph of $y = x^2$ about the x-axis.

The absolute value function, `abs(`, is found in the `NUM` submenu of [MATH].

Enter the equation $y = |x|$ into the function editor.

[Y=] [CLEAR] [MATH] [▶] [1] [X,T,Θ,n] [)] [ENTER] [CLEAR]

Plot1 Plot2 Plot3
\Y1=abs(X)
\Y2=
\Y3=
\Y4=
\Y5=
\Y6=
\Y7=

Graph $y = |x|$.

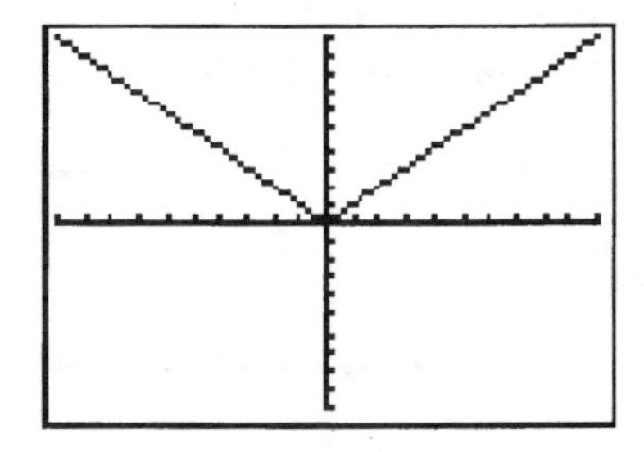

Return to the function editor and enter $y = -|x|$ into y2 and graph the two equations.

[Y=] [▼] [(-)] [MATH] [▶] [1] [X,T,Θ,n] [)] [ENTER]

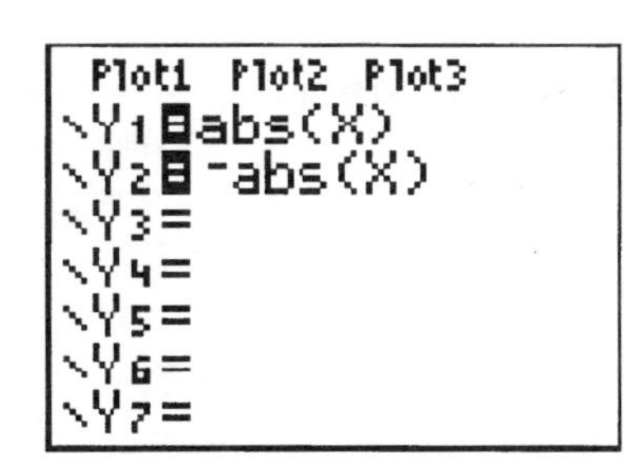

Graph the two equations.

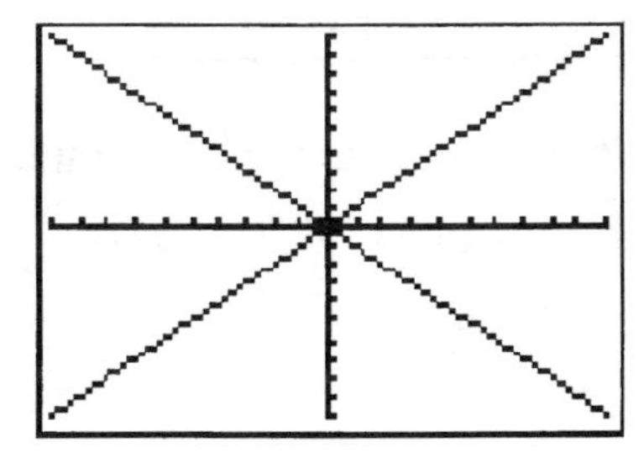

Notice that the new graph is another v shaped graph, the only difference is that the v shape is on the opposite side of the x-axis, and it opens down. In other words, the graph of $y=-|x|$ is obtained by reflecting the graph of $y=|x|$ about the x-axis.

If we multiply the formula for a function by -1, the resulting graph is a reflection of the graph of the given function about the x-axis.

49. **Exploration** Graph $y=x^3$. Then on the same screen graph $y=(x-1)^3+2$. Could you have predicted the result?

Enter $y=x^3$ in the function editor. Go to `WINDOW` and enter limits for x and y. Let's use the standard window $-10 \le x \le 10$ and $-10 \le y \le 10$.

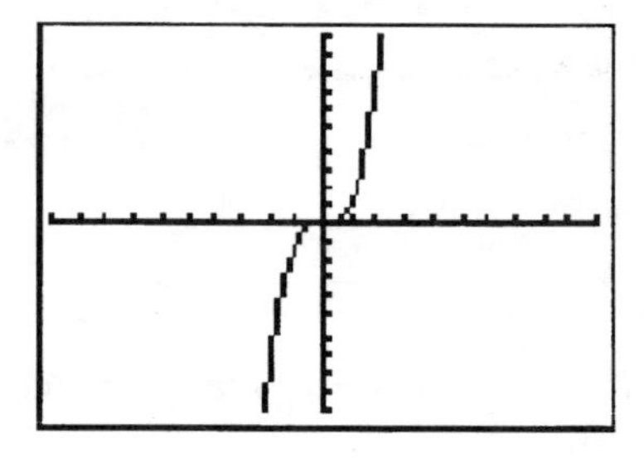

Return to the function editor and enter $y=(x-1)^3+2$ into `y2` and graph the two equations.

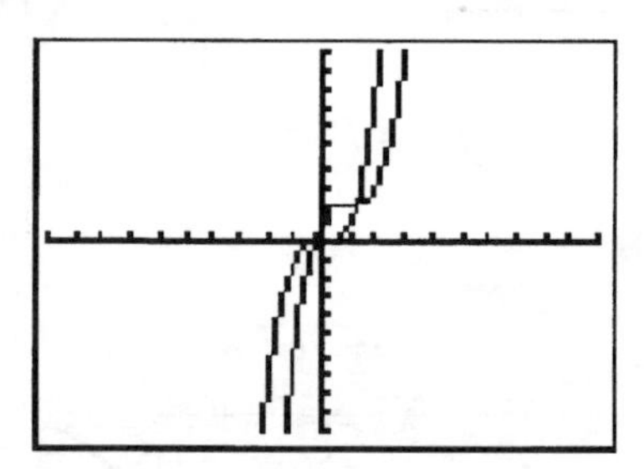

Yes, based on the results of Problems 45 and 46, we could determine that the graph of $y=(x-1)^3+2$ is obtained from the graph of $y=x^3$ by shifting it to the right one unit and up two units.

51. **Exploration** Graph $y = x^3$, $y = x^5$, and $y = x^7$ on the same screen. What do you notice is the same about each graph? What do you notice that is different?

Enter $y = x^3$, $y = x^5$, and $y = x^7$ in the function editor. Go to `WINDOW` and enter limits for x and y. Let's use the window $-2 \le x \le 2$ and $-2 \le y \le 2$.

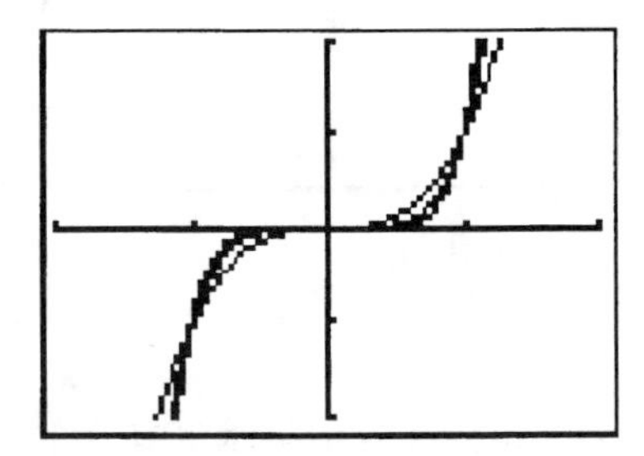

All three graphs have the same general shape and all three graphs pass through $(0,0)$, $(1,1)$, and $(-1,-1)$.

As the degree increases, the graph rises faster for values of x that are greater than 1, and falls faster for values of x that are less than -1. Also, as the degree increases, the graph is closer to the x-axis for values of x that are between -1 and 1.

Section 10.5 Graphing Techniques: Shifts and Reflections

41. **Exploration**

(a) Use a graphing utility to graph $y = x+1$ and $y = |x+1|$.

Enter $y = x+1$ in the function editor. Go to `WINDOW` and enter limits for x and y. Let's use the standard window $-10 \le x \le 10$ and $-10 \le y \le 10$.

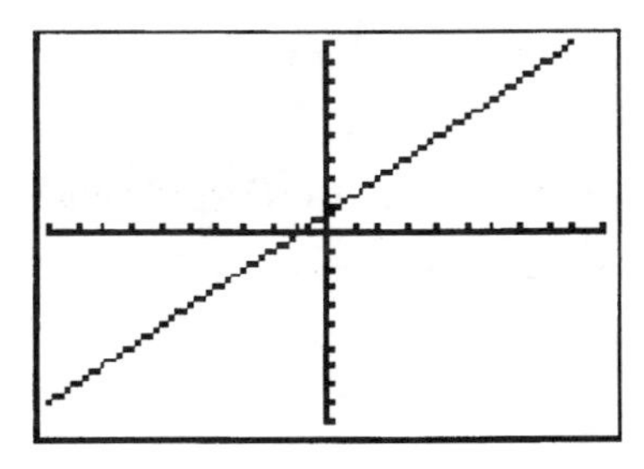

Return to the function editor and enter $y = |x+1|$ into y1 and graph the equation.

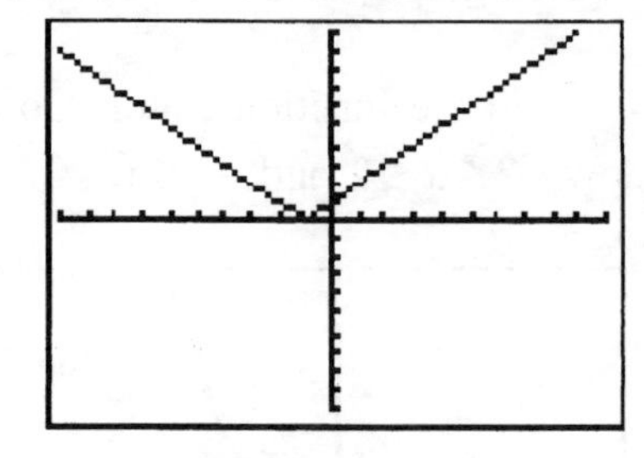

(b) Graph $y = 4 - x^2$ and $y = |4 - x^2|$.

Enter $y = 4 - x^2$ in the function editor. Go to WINDOW and enter limits for x and y. Let's use the standard window $-10 \le x \le 10$ and $-10 \le y \le 10$.

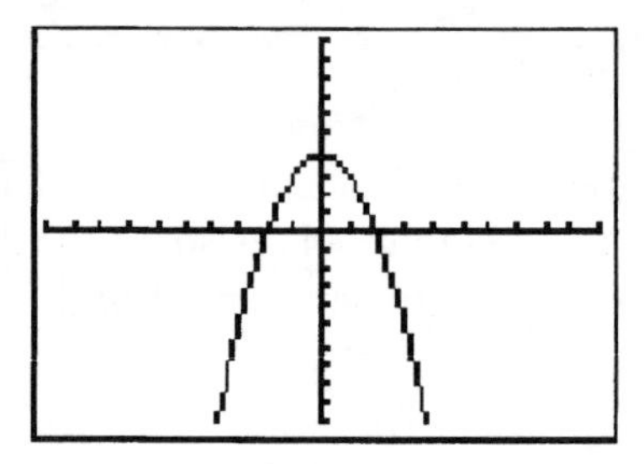

Return to the function editor and enter $y = |4 - x^2|$ into y1 and graph the equation.

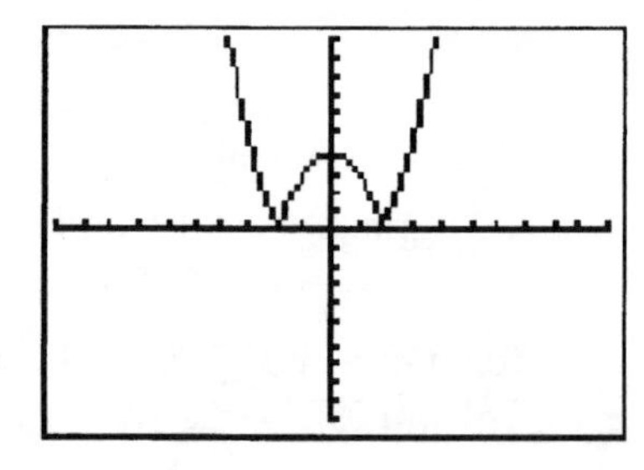

(c) Graph $y = x^3 + x$ and $y = |x^3 + x|$.

Enter $y = x^3 + x$ in the function editor. Go to WINDOW and enter limits for x and y. Let's use the standard window $-10 \le x \le 10$ and $-10 \le y \le 10$.

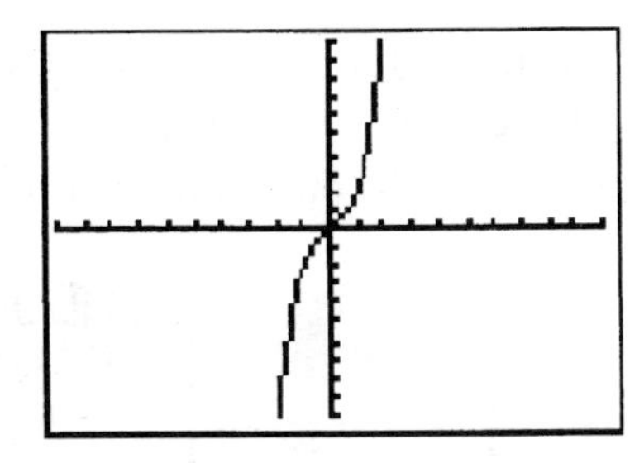

Return to the function editor and enter $y = |x^3 + x|$ into y1 and graph the equation.

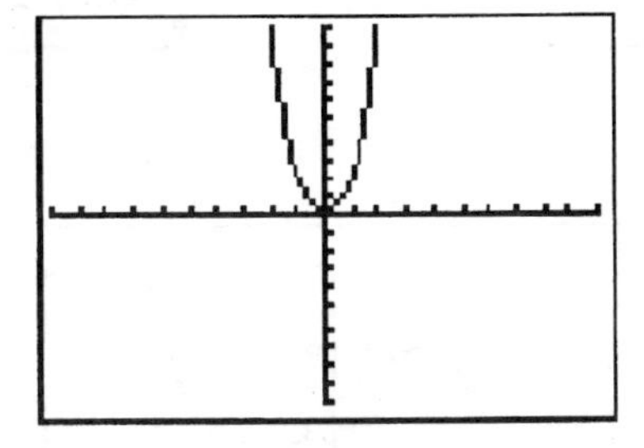

(d) What do you conclude about the relationship between the graphs of $y = f(x)$ and $y = |f(x)|$?

The graphs of $y = f(x)$ and $y = |f(x)|$ are identical on the x-intervals where $f(x) \geq 0$. However, for any x-interval where $f(x) < 0$, the graph of $y = |f(x)|$ is a reflection of the graph of $y = f(x)$ about the x-axis.

Chapter 10 Review

In Problems 39–42, use a graphing utility to graph each function over the indicated interval. Approximate any local maxima and local minima. Determine where the function is increasing and where it is decreasing.

39. $f(x) = 2x^3 - 5x + 1 \quad (-3, 3)$

Enter the formula for f in the function editor. Go to WINDOW and enter limits for x. We must determine the limits for y so let's try $-10 \le y \le 10$. If this does not work, then we can go back to WINDOW and adjust the limits on y until we find a good window.

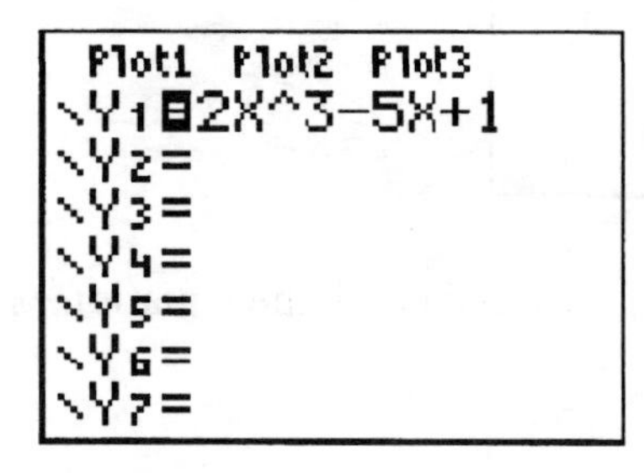

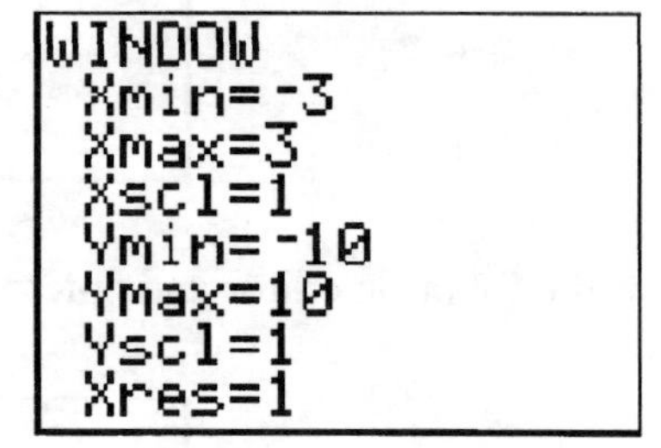

Graph the function.

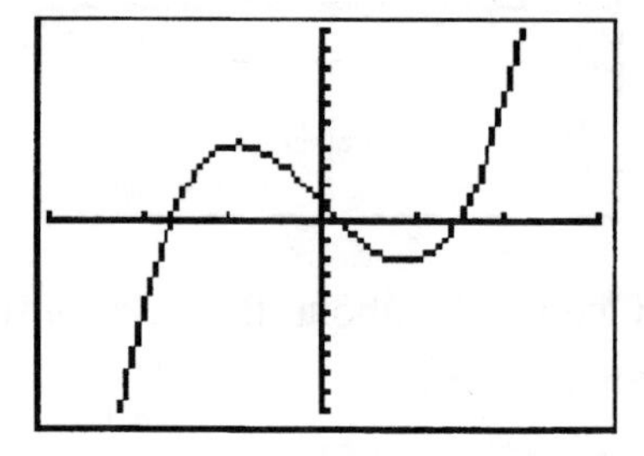

This window is suitable.

Use maximum to find the local maximum. Note that the x-coordinate of the local maximum is between $x = -2$ and $x = 0$.

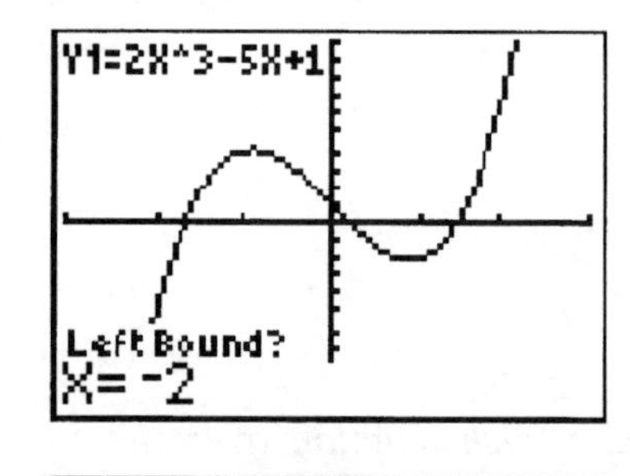

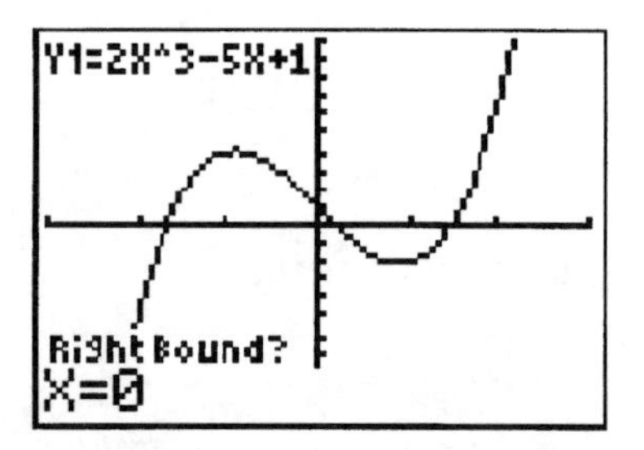

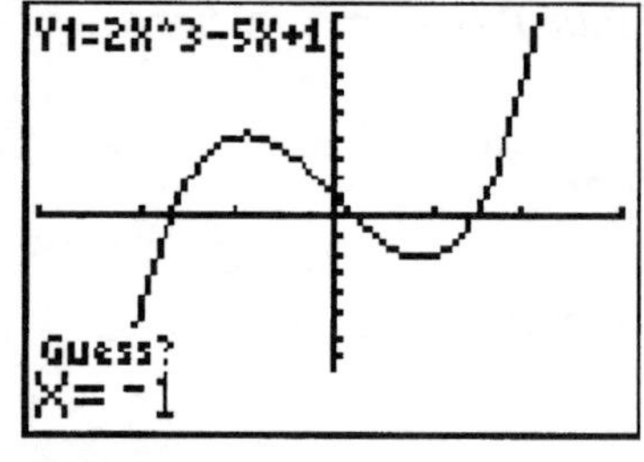

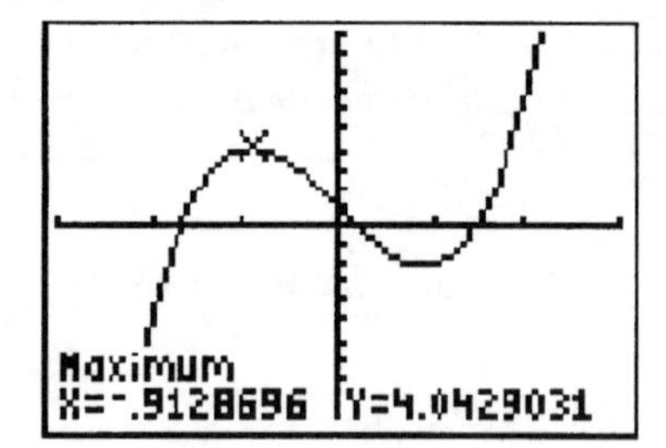

The local maximum is approximately $(-0.913, 4.043)$.

Use `minimum` to find the local minimum. Note that the x-coordinate of the local minimum is between $x = 0$ and $x = 2$.

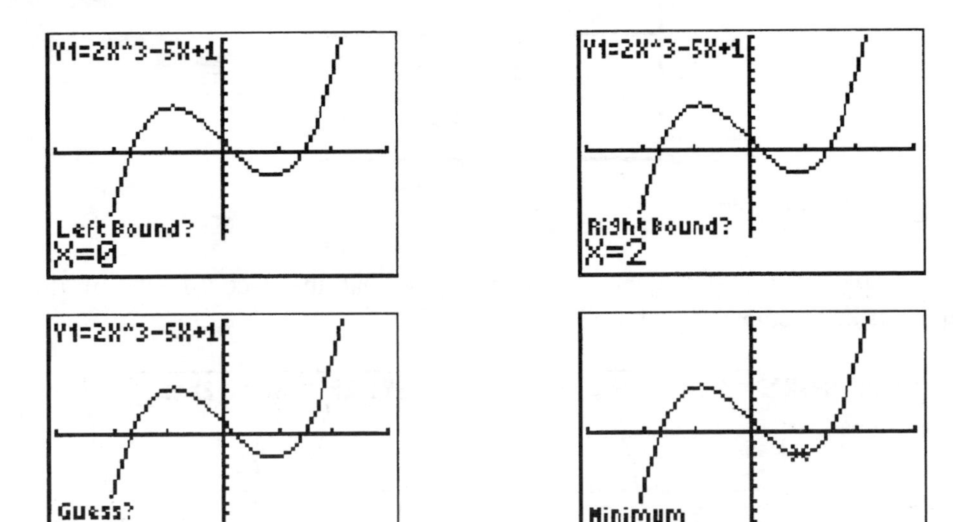

The local minimum is approximately $(0.913, -2.043)$.

Thus, f is increasing on the intervals $(-3, -0.913)$ and $(0.913, 3)$, and f is decreasing on the interval $(-0.913, 0.913)$.

41. $f(x) = 2x^4 - 5x^3 + 2x + 1 \quad (-2, 3)$

Enter the formula for f in the function editor. Go to `WINDOW` and enter limits for x. We must determine limits for y, so try $-10 \le y \le 10$. If this does not work, then we can go back to `WINDOW` and adjust the limits on y until we find a good window.

```
Plot1 Plot2 Plot3
\Y1=2X^4-5X^3+2X
+1
\Y2=
\Y3=
\Y4=
\Y5=
\Y6=
```

```
WINDOW
 Xmin=-2
 Xmax=3
 Xscl=1
 Ymin=-10
 Ymax=10
 Yscl=1
 Xres=1
```

Graph the function.

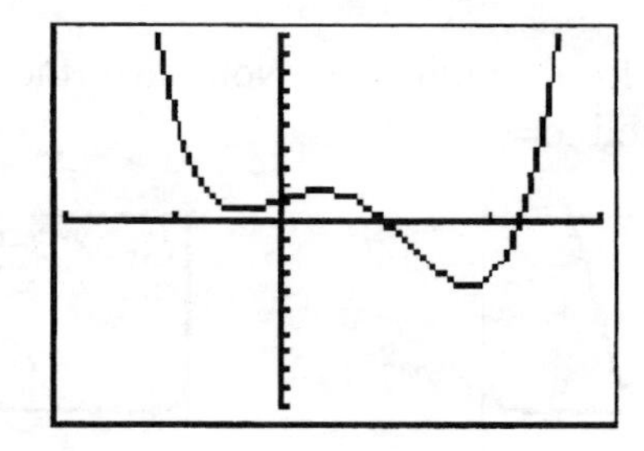

This window is suitable.

Use `maximum` to find the local maximum. Note that the x-coordinate of the local maximum is between $x=0$ and $x=1$.

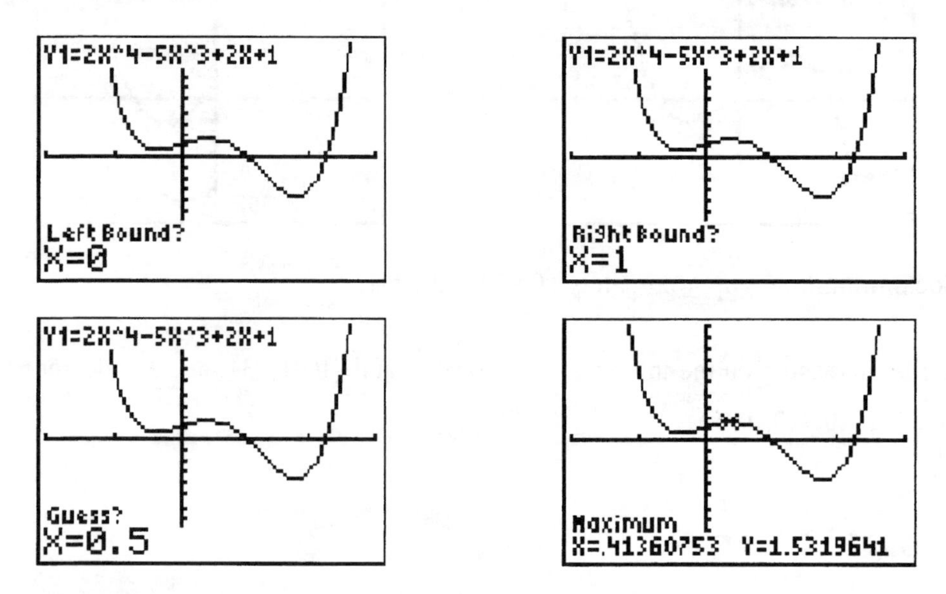

The local maximum is approximately $(0.414, 1.532)$.

Use `minimum` to find both local minima. Note that the x-coordinate of the first local minimum is between $x=-1$ and $x=0$.

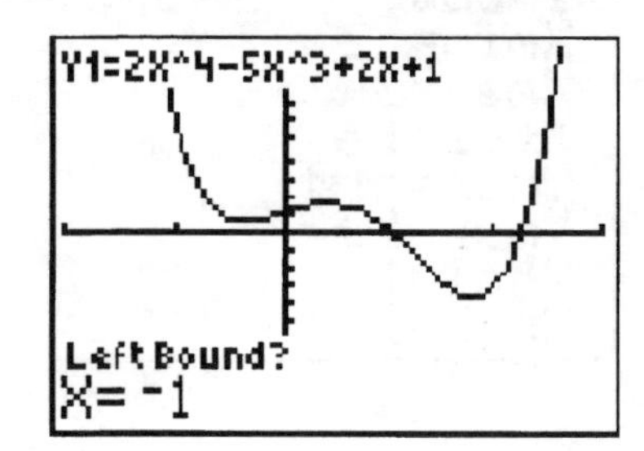

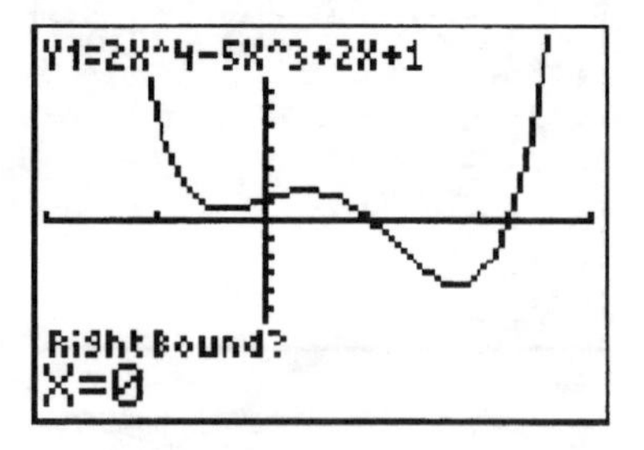

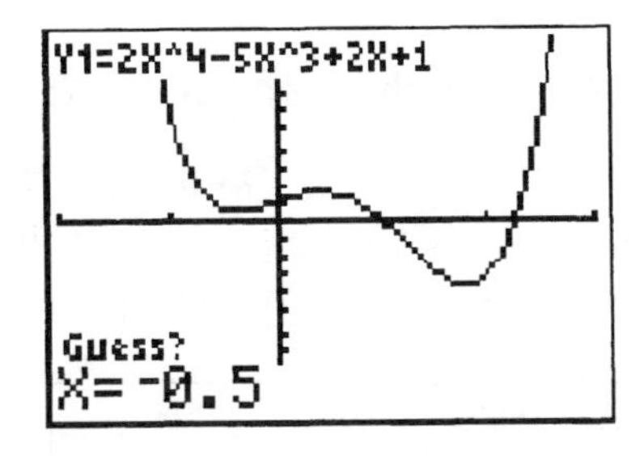

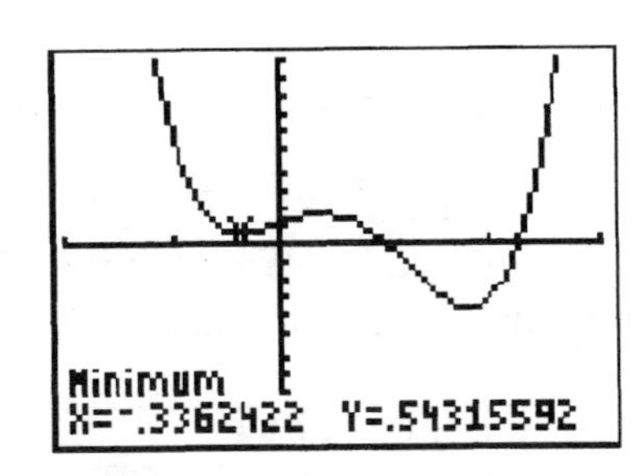

The first local minimum is approximately $(-0.336, 0.543)$.

The x-coordinate of the second local minimum is between $x=1$ and $x=3$.

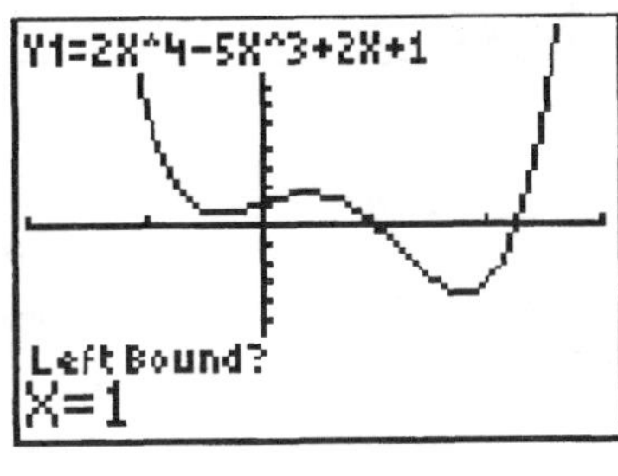

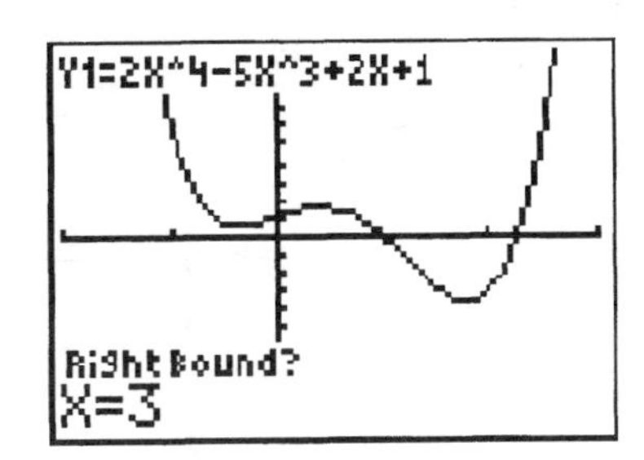

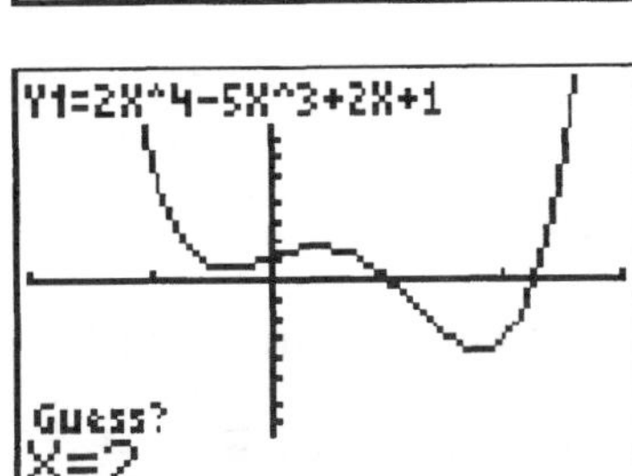

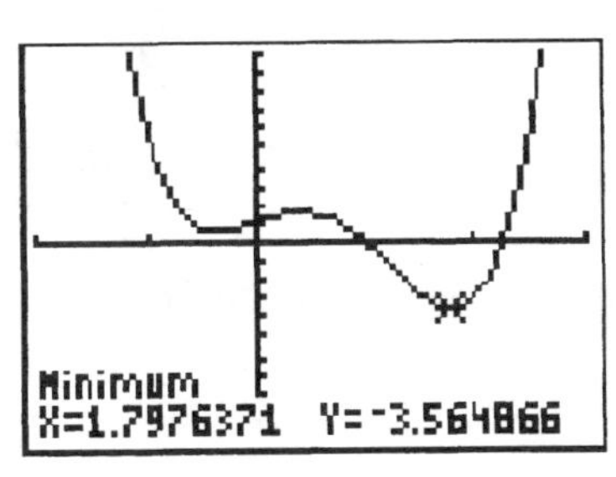

The second local minimum is approximately $(1.798, -3.565)$.

Thus, f is increasing on the intervals $(-0.336, 0.414)$ and $(1.798, 3)$, and f is decreasing on the intervals $(-2, -0.336)$ and $(0.414, 1.798)$.

77. **Cost of a Drum** A drum in the shape of a right circular cylinder is required to have a volume of 500 cubic centimeters. The top and bottom are made of material that costs 6¢ per square centimeter; the sides are made of material that costs 4¢ per square centimeter. Hint: The volume V of a right circular cylinder of height h and radius r is $V = \pi r^2 h$.

    (d) Graph $C = C(r)$. For what value of r is the cost C least?

The cost function is given by $C(r) = 0.12\pi r^2 + \frac{40}{r}$. Enter the formula for C in the function editor. Be sure to use X for r when you enter the equation. Go to WINDOW and enter limits for x and y. We are not given any limits for either variable, although for this problem to make sense we must have $x > 0$. Let's try the following window.

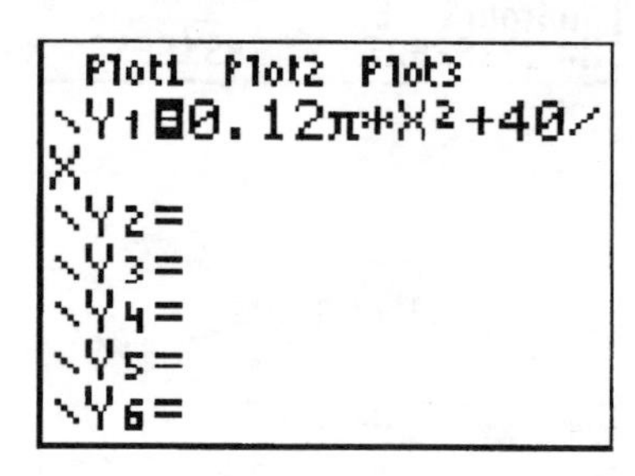

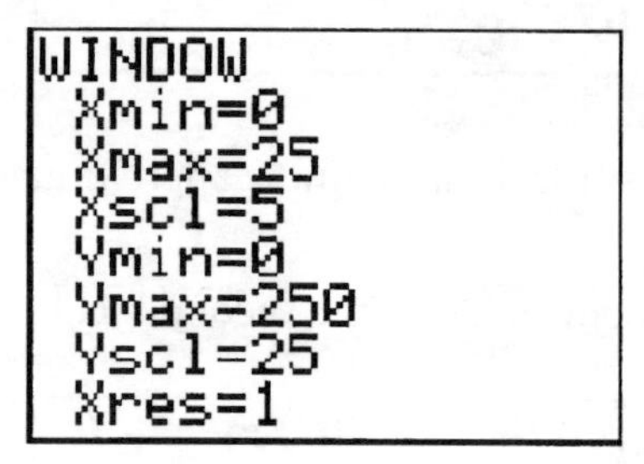

Graph C.

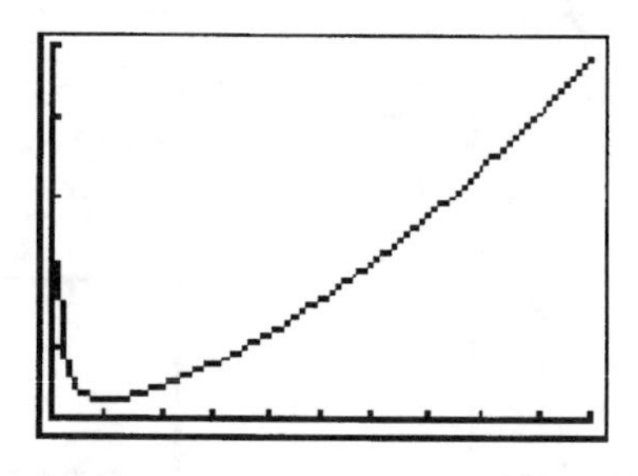

Find the local minimum. Notice that the minimum is between $x = 1$ and $x = 10$.

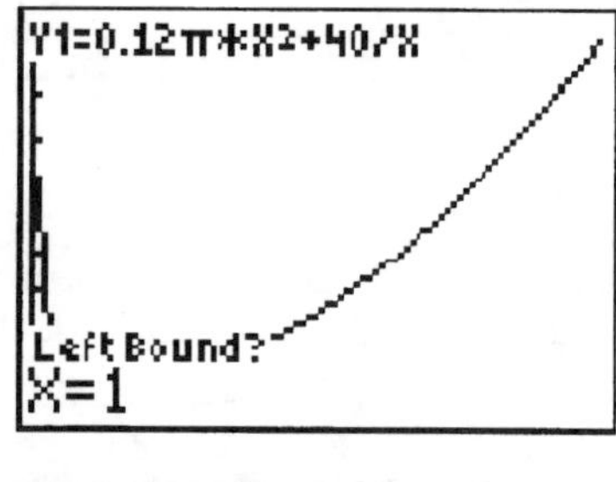

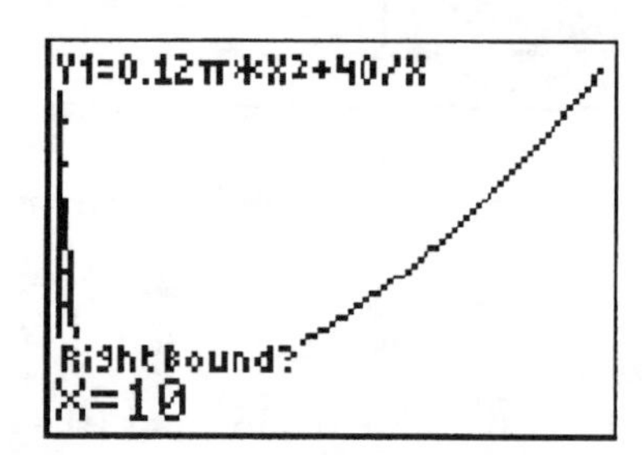

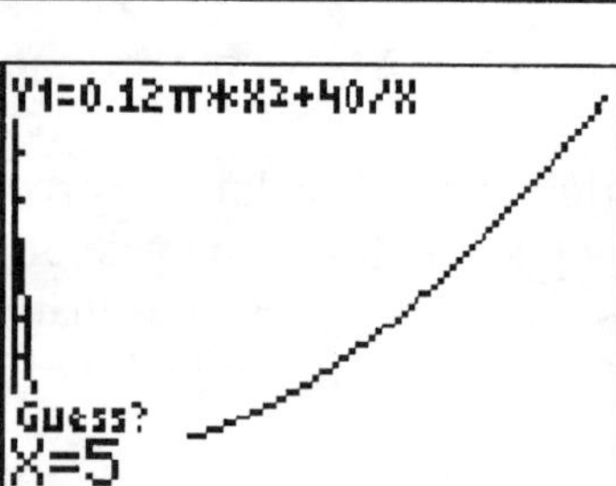

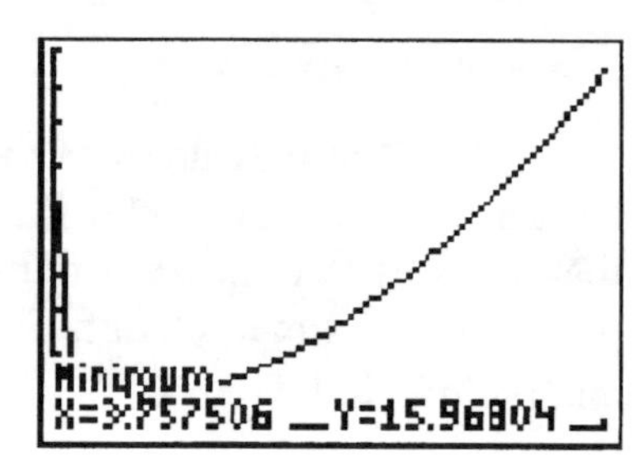

The local minimum is approximately $(3.76, 15.97)$. If r is approximately 3.76 centimeters, then the cost C will be minimized.

Summary

The commands introduced in this chapter are:

`abs(`*value*`)`

`intersect`

`maximum`

`minimum`

`Table`

Chapter 11 – Classes of Functions

Section 11.1 Quadratic Functions

47. **Analyzing the Motion of a Projectile** A projectile is fired from a cliff 200 feet above the water at an inclination of 45° to the horizontal, with a muzzle velocity of 50 feet per second. The height h of the projectile above the water is given by

$$h(x) = \frac{-32x^2}{(50)^2} + x + 200$$

where x is the horizontal distance of the projectile from the base of the cliff.

(d) Using a graphing utility, graph the function h, $0 \le x \le 200$.

Enter the formula for h into the function editor. Go to WINDOW and enter limits for x. We must determine limits for y, so try $0 \le y \le 300$. If this does not work, then we can go back to WINDOW and adjust the limits on y until we find a good window.

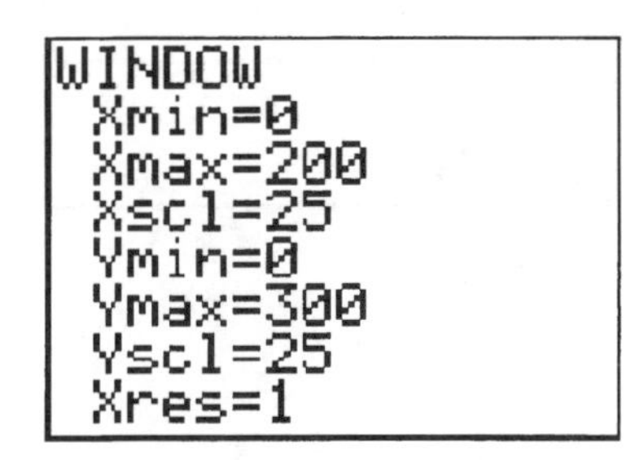

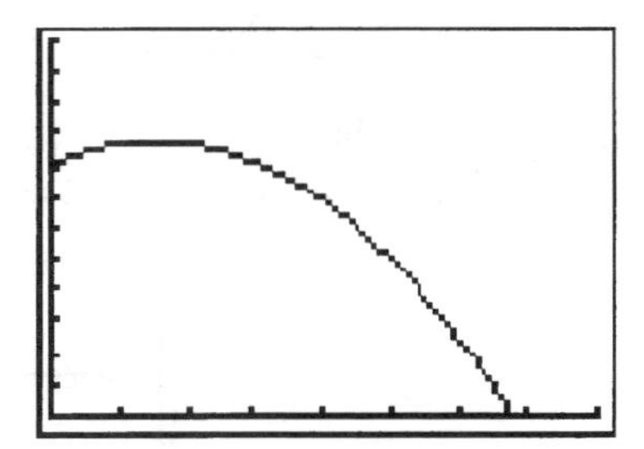

(e) When the height of he projectile is 100 feet above the water, how far is it from the cliff?

Return to the function editor, input the equation $y = 100$, and graph the two equations.

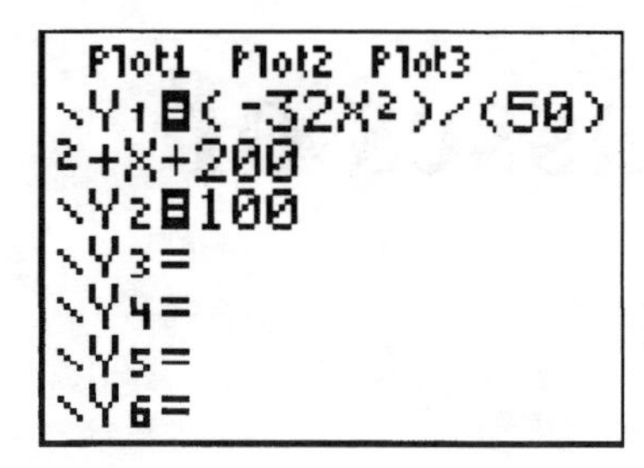

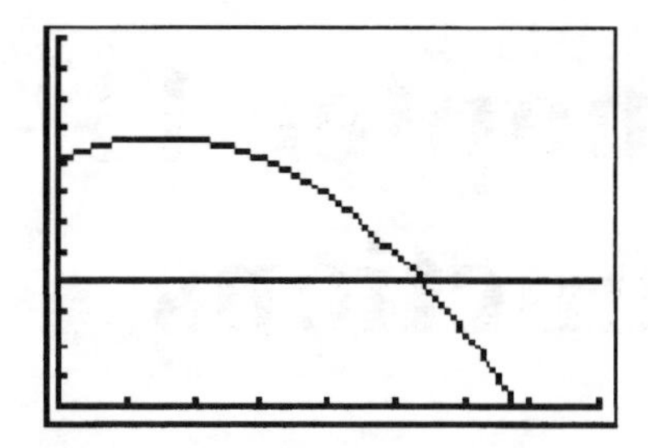

The point where the graph of h and the horizontal line $y = 100$ intersect represents the point where the projectile is 100 feet above the water. Recall that you can find the point of intersection on your calculator using the `intersect` function, which can be found in the CALC menu. The x-coordinate of the point of intersection appears to be close to $x = 125$.

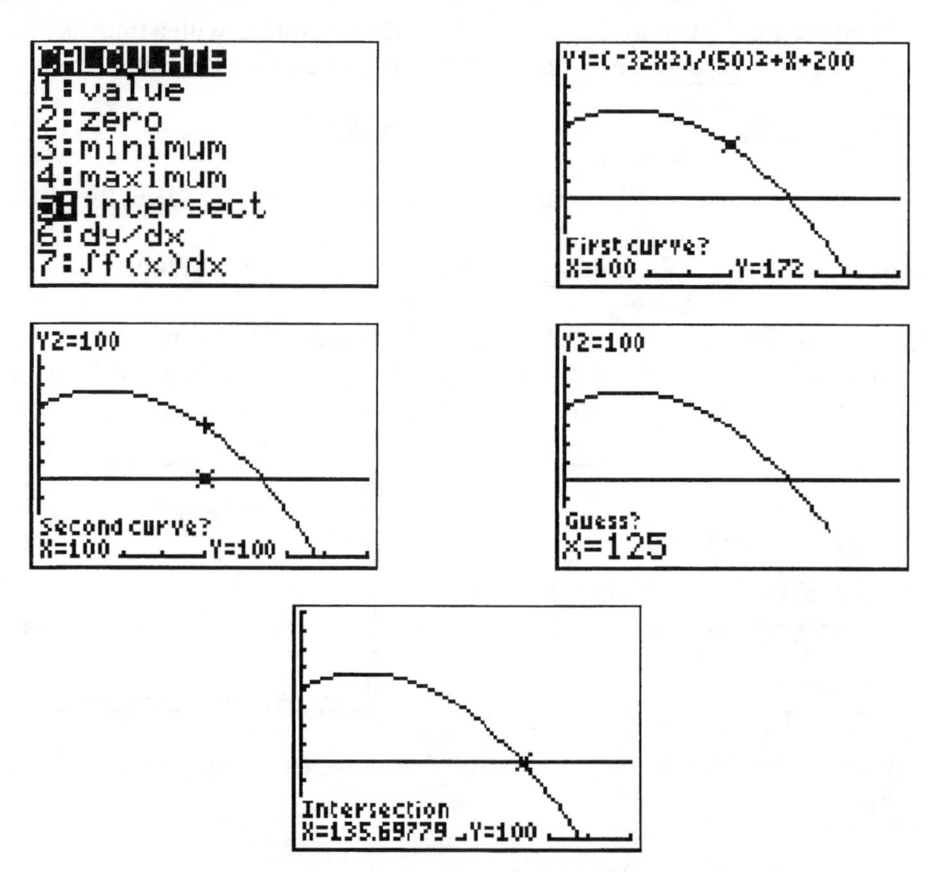

Thus, the point of intersection is approximately $(135.698, 100)$. When the projectile is 100 feet above the water it is approximately 135.698 feet from the cliff.

53. **Hunting** The function $H(x) = -1.01x^2 + 114.3x + 451.0$ models the number of individuals who engage in hunting activities whose annual income is x thousand dollars.

(b) Using a graphing utility, graph $H = H(x)$. Are the number of hunters increasing or decreasing for individuals earning between \$20,000 and \$40,000?

Enter the formula for H into the function editor. Since x is in thousands, the earnings range of \$20,000 to \$40,000 corresponds to $20 \le x \le 40$. Go to `WINDOW` and enter limits $20 \le x \le 40$ for x. We must determine limits for y, so try $0 \le y \le 1000$. If this does not work, then we can go back to `WINDOW` and adjust the limits on y until we find a good window.

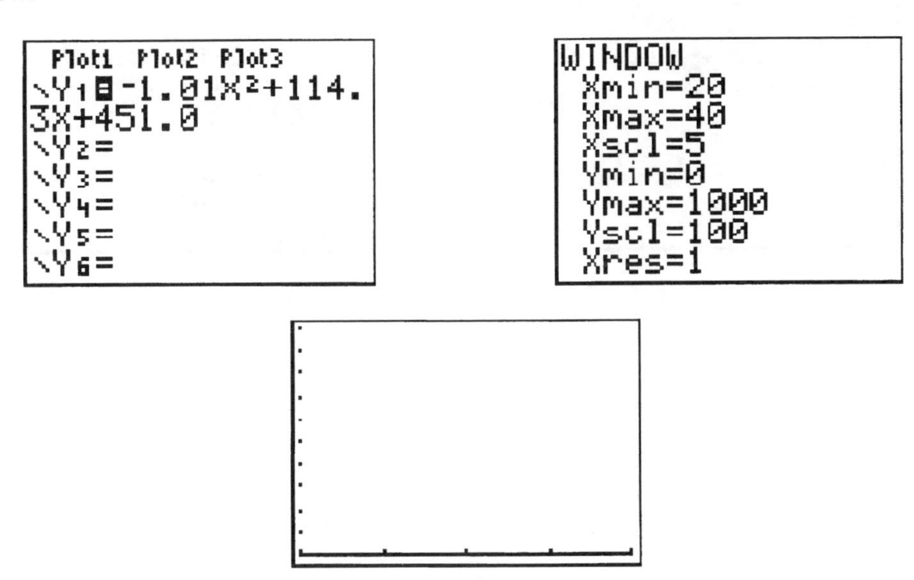

There is no graph in our viewing rectangle, so we will need to find a better window. Lets try $0 \le y \le 5000$.

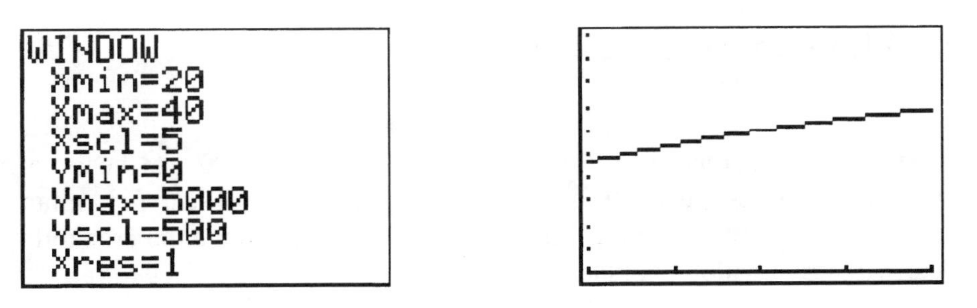

This time we can see the graph. The graph is increasing on this interval, so the number of hunters is increasing for individuals earning between $20,000 and $40,000.

55. **Male Murder Victims** The function $M(x) = 0.76x^2 - 107.00x + 3854.18$ models the number of male murder victims who are x years of age ($20 \le x \le 90$).

(c) Using a graphing utility, graph $M = M(x)$.

Enter the formula for h into the function editor. Go to WINDOW and enter limits for x. We must determine limits for y, so try $0 \le y \le 4000$. If this does not work, then we can go back to WINDOW and adjust the limits on y until we find a good window.

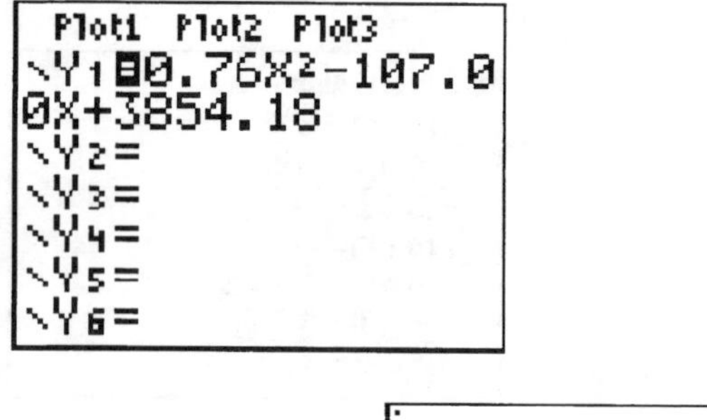

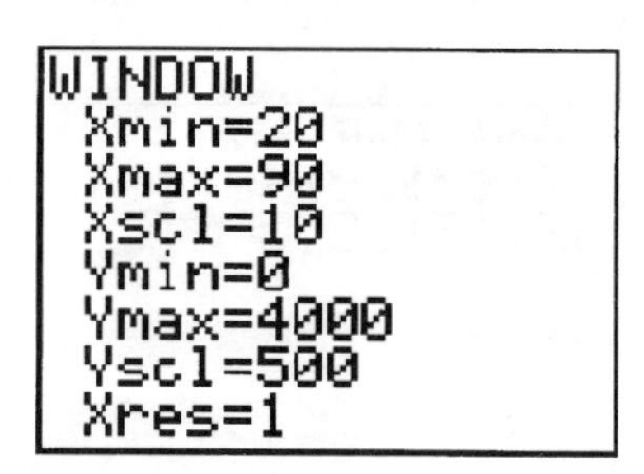

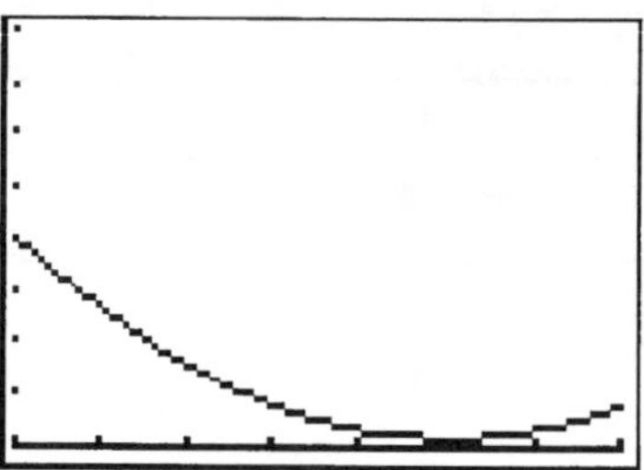

Section 11.3 Exponential Functions

63. **Exponential Probability** Between 12:00 PM and 1:00 PM, cars arrive at Citibank's drive-thru at the rate of 6 cars per hour (0.1 car per minute). The following formula from probability can be used to determine the probability that a car will arrive within t minutes of 12:00 PM:

$$F(t) = 1 - e^{-0.1t}$$

(d) Graph F using your graphing utility.

Enter the formula for F into the function editor using x instead of t. Since x is in minutes, the time range 12:00 PM to 1:00 PM corresponds to $0 \le x \le 60$. Go to WINDOW and enter the limits $0 \le x \le 60$ for x. We must determine limits for y, so try $0 \le y \le 1$. If this does not work, then we can go back to WINDOW and adjust the limits on y until we find a good window.

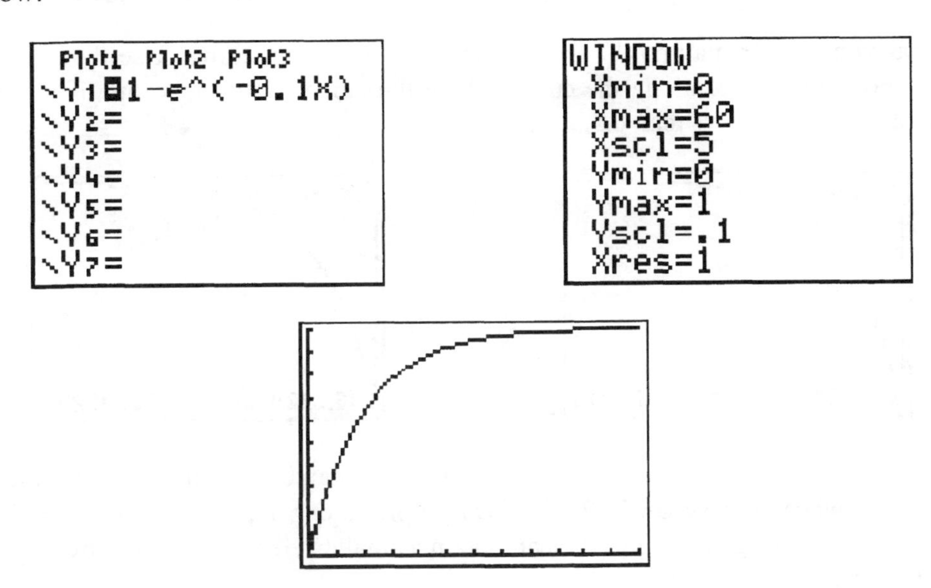

(e) Using TRACE, determine how many minutes are needed for the probability to reach 50%.

The TRACE feature uses the points on the curve that the calculator generated to draw the curve. As you TRACE along the curve, the distance between successive x-coordinates of points is the same. It is determined by

$$\Delta x = \frac{\text{Xmax} - \text{Xmin}}{94}$$

Because of this, as we trace along the curve, we may not get "nice" values for x (with the window used for the problem we will not get nice values). We must trace until we find the value of x that corresponds to the value y that is closest to 0.5.

Select the TRACE feature.

[TRACE]

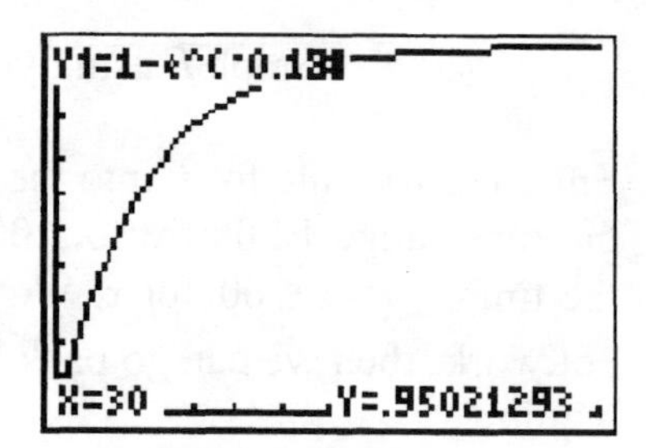

Note that your cursor may not be in the same spot as shown above. Use the [◄] and [►] keys to move the cursor along the curve. Move the cursor until the value of y changes from more than 0.5 to less than 0.5.

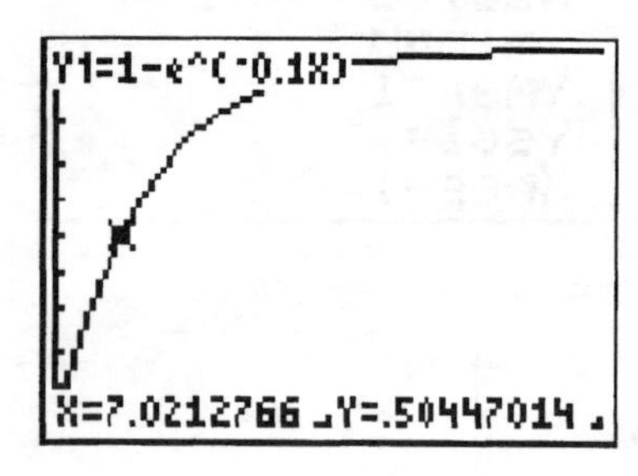

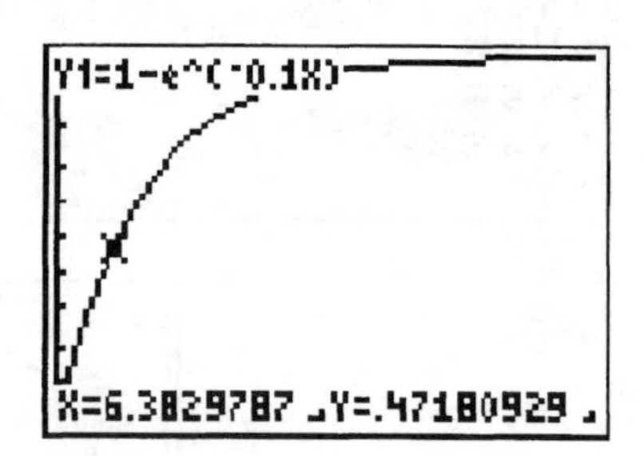

Since y is less than 0.5 when $x \approx 6.383$, and y is more than 0.5 when $x \approx 7.021$, then y is equal to 0.5 when $6.383 < x < 7.021$. We can improve our approximation if we TRACE over a smaller viewing rectangle. We can obtain a smaller viewing window by "zooming in" on the graph using Zoom In.

[ZOOM]

Select Zoom In.

[ENTER]

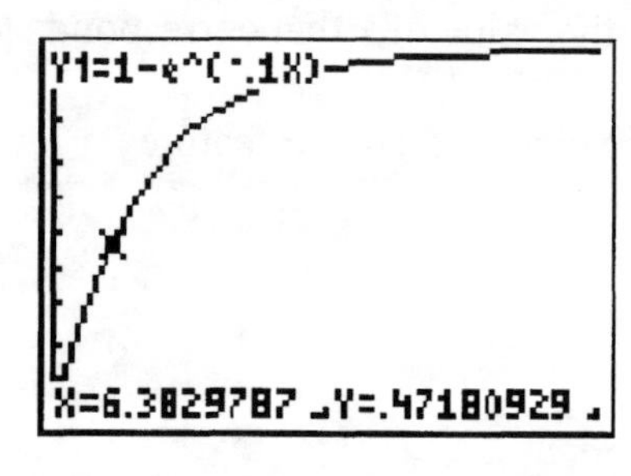

Be sure the y-value of the cursor is close to 0.5. Zoom in on the graph.

[2]

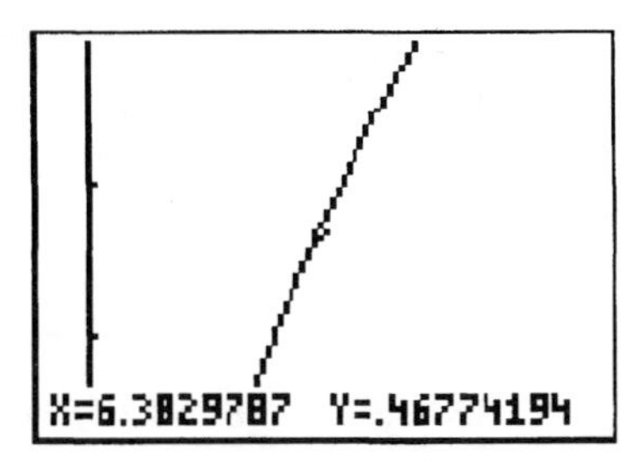

Use TRACE and move the cursor so that the y-value is close 0.5.

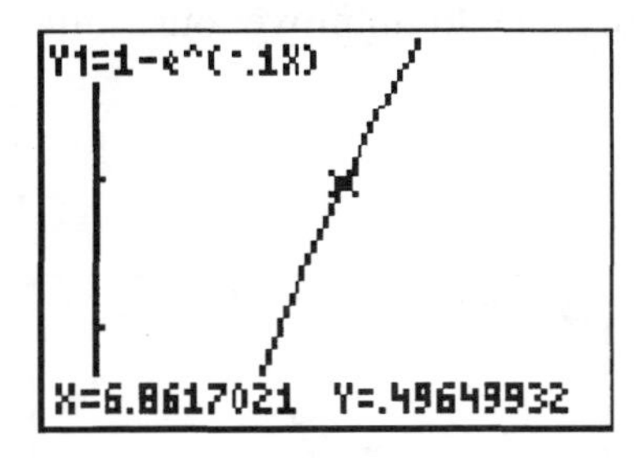

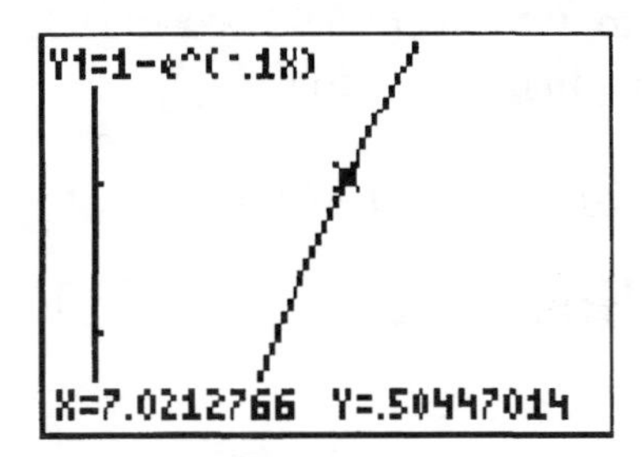

Since y is less than 0.5 when $x \approx 6.861$, and y is more than 0.5 when $x \approx 7.021$, then y is equal to 0.5 when $6.383 < x < 7.021$. We can continue to improve our solution by zooming in on our graph.

Zoom In again, then TRACE.

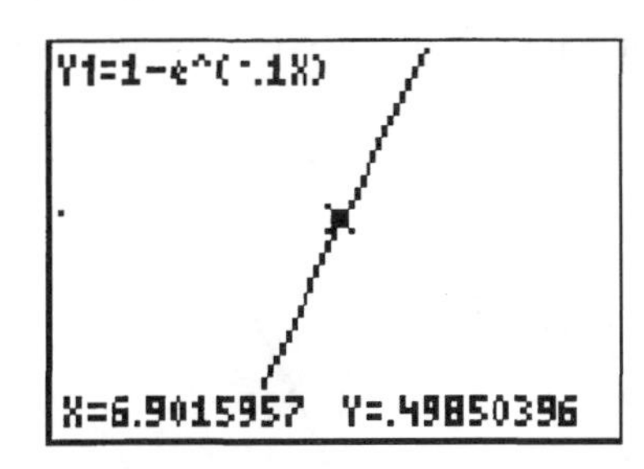

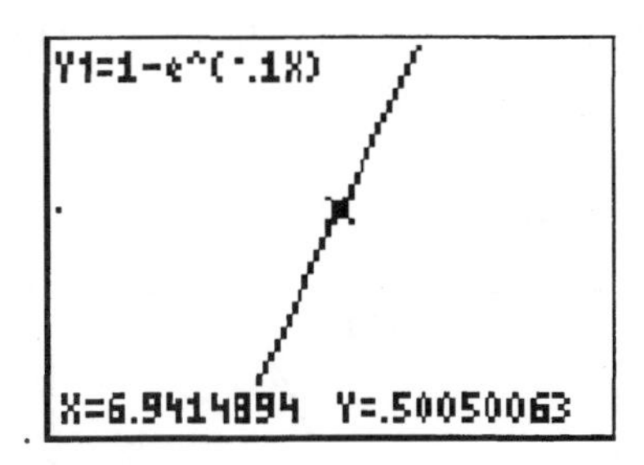

Since y is less than 0.5 when $x \approx 6.902$, and y is more than 0.5 when $x \approx 6.914$, then y is equal to 0.5 when $6.902 < x < 6.914$. We can continue to improve our solution by zooming in on our graph.

`Zoom In` again, then TRACE.

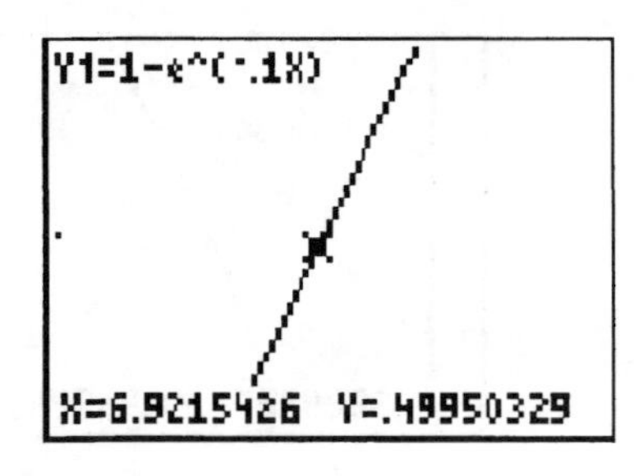

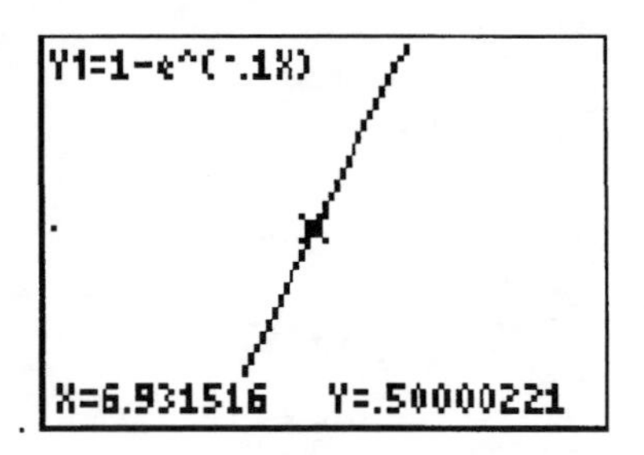

Since y is less than 0.5 when $x \approx 6.921$, and y is more than 0.5 when $x \approx 6.932$, then y is equal to 0.5 when $6.921 < x < 6.932$. We can continue to improve our solution by zooming in on our graph.

`Zoom In` again, then TRACE.

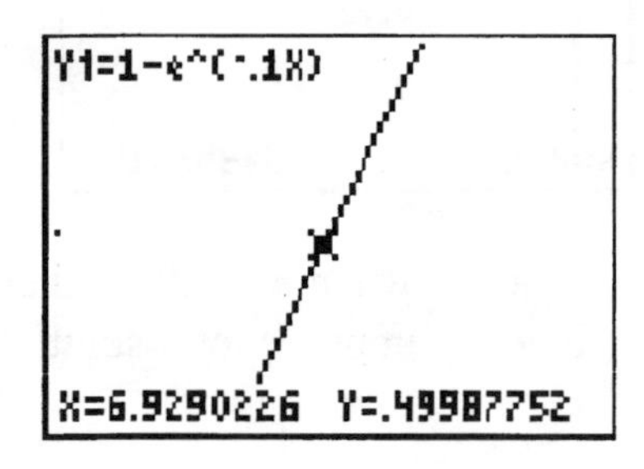

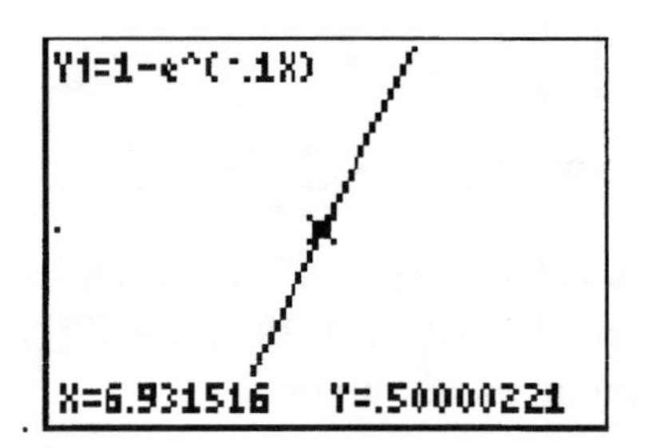

Since y is less than 0.5 when $x \approx 6.929$, and y is more than 0.5 when $x \approx 6.932$, then y is equal to 0.5 when $6.929 < x < 6.932$. We can continue to improve our solution by zooming in on our graph.

`Zoom In` again, then TRACE.

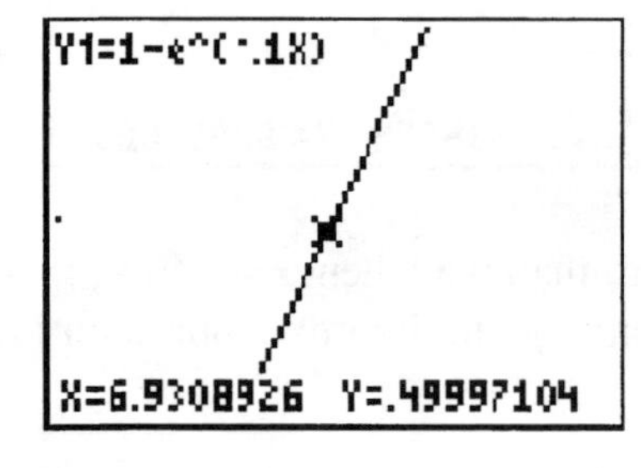

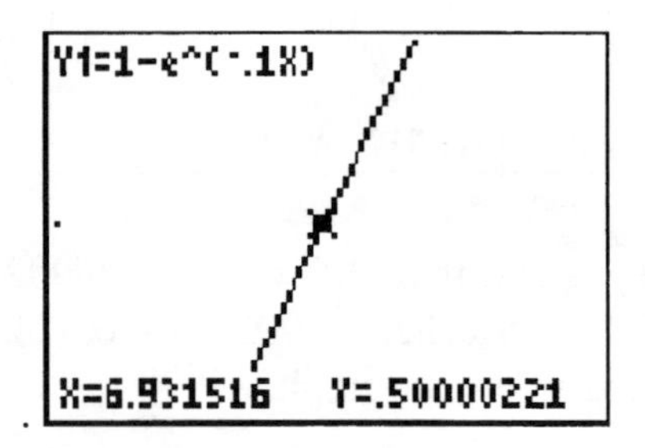

Since y is less than 0.5 when $x \approx 6.9309$, and y is more than 0.5 when $x \approx 6.9315$, then y is 0.5 when x is approximately 6.931. Thus, the probability will reach 50% after 6.931 minutes.

79. The **hyperbolic sine function**, designated by sinh x, is defined as

$$\sinh x = \frac{1}{2}\left(e^{x} - e^{-x}\right)$$

(b) Graph $f(x) = \sinh x$ using a graphing utility.

Enter the formula for f into the function editor. Go to WINDOW and enter limits for x and y. Since we are not given limits for x or y, use the standard window, $-10 \le x \le 10$ and $-10 \le y \le 10$.

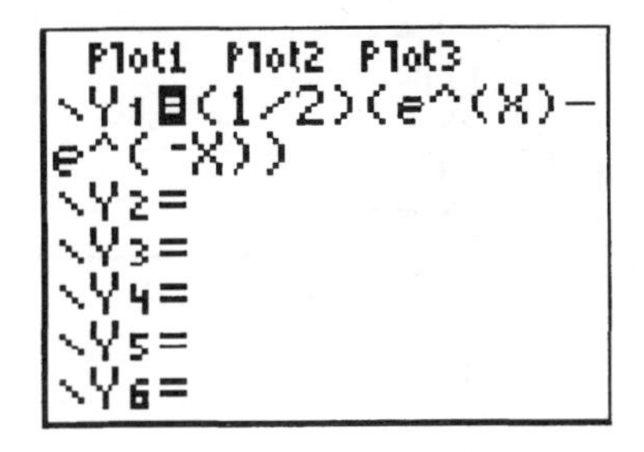

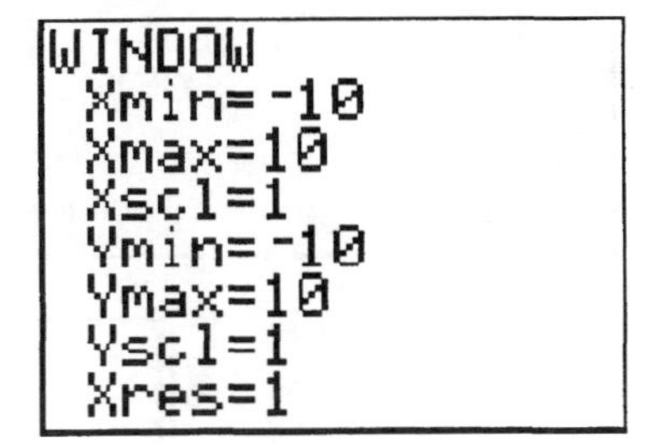

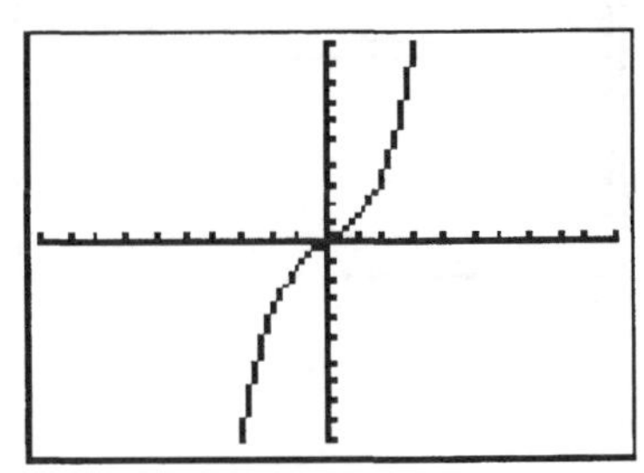

Section 11.5 Properties of Logarithms

In Problems 67–72, graph each function using a graphing utility and the Change-of-Base formula.

67. $y = \log_4 x$

Recall that the Change-of-Base formula states that $\log_a M = \dfrac{\log_b M}{\log_b a}$. Because we will use our calculator to graph the equation, we must use 10 or e for the new base b. Use e as the new base. Thus

$$y = \log_4 x = \frac{\log_e x}{\log_e 4} = \frac{\ln x}{\ln 4}$$

Enter $y = \frac{\ln x}{\ln 4}$ into the function editor. Go to WINDOW and enter limits for x and y. Since we are not given limits for x or y, use the viewing rectangle $0 \le x \le 10$ and $-10 \le y \le 10$.

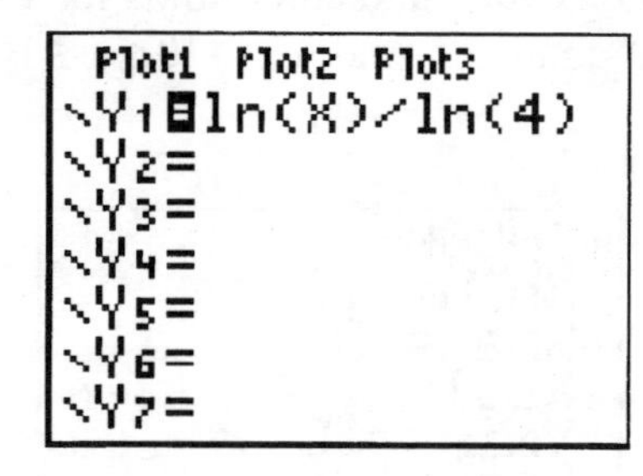

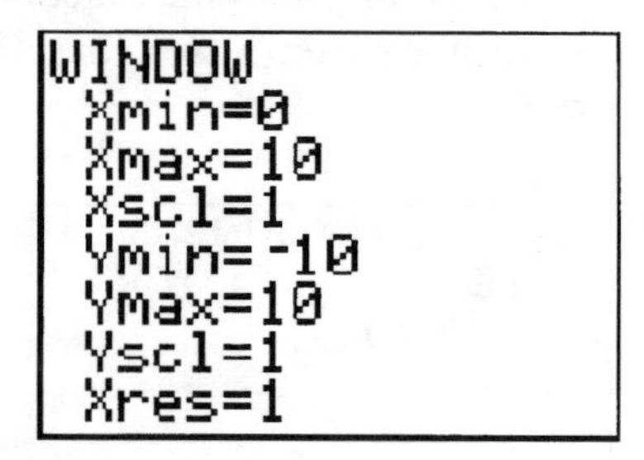

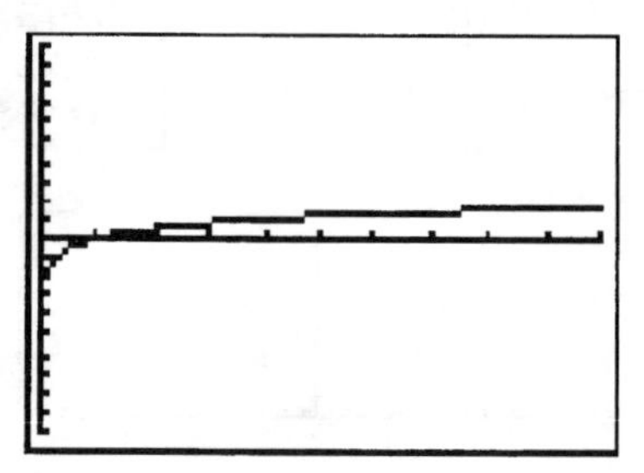

69. $y = \log_2 (x+2)$

Using the Change-of-Base formula we obtain

$$y = \log_2 (x+2) = \frac{\ln(x+2)}{\ln 2}$$

Enter $y = \frac{\ln(x+2)}{\ln 2}$ into the function editor. Go to WINDOW and enter limits for x and y. Since we are not given limits for x or y, use the viewing rectangle $-3 \le x \le 10$ and $-10 \le y \le 10$.

```
Plot1 Plot2 Plot3
\Y1■ln(X+2)/ln(2
)
\Y2=
\Y3=
\Y4=
\Y5=
\Y6=
```

```
WINDOW
 Xmin=-3
 Xmax=10
 Xscl=1
 Ymin=-10
 Ymax=10
 Yscl=1
 Xres=1
```

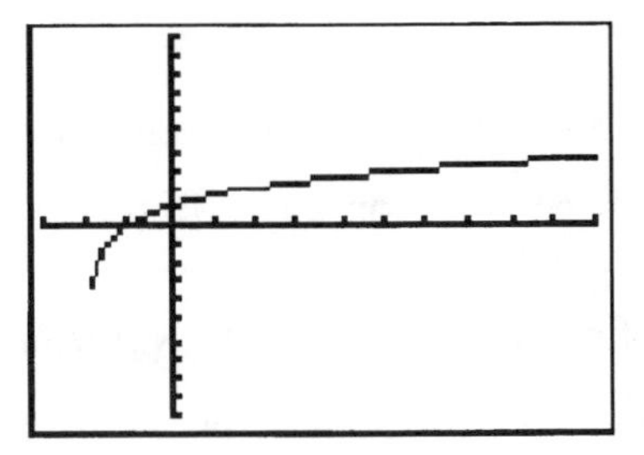

71. $y = \log_{x-1}(x+1)$

Using the Change-of-Base formula we obtain

$$y = \log_{x-1}(x+1) = \frac{\ln(x+1)}{\ln(x-1)}$$

Enter $y = \dfrac{\ln(x+1)}{\ln(x-1)}$ into the function editor. Go to WINDOW and enter limits for x and y.
Since we are not given limits for x or y, use the viewing rectangle $0 \le x \le 10$ and $-10 \le y \le 10$.

```
Plot1 Plot2 Plot3
\Y1■ln(X+1)/ln(X
-1)
\Y2=
\Y3=
\Y4=
\Y5=
\Y6=
```

```
WINDOW
 Xmin=0
 Xmax=10
 Xscl=1
 Ymin=-10
 Ymax=10
 Yscl=1
 Xres=1
```

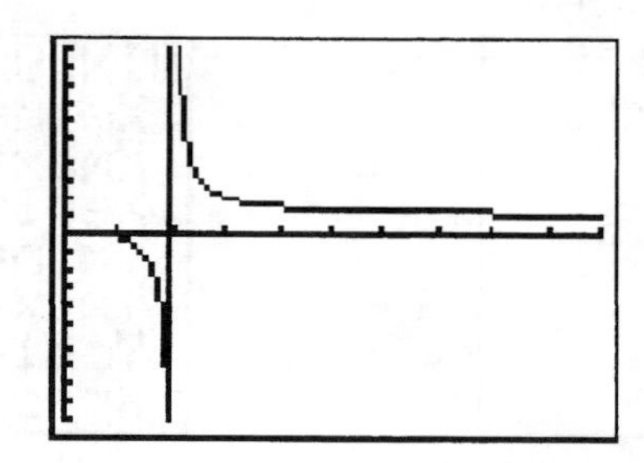

97. Graph $Y_1 = \log(x^2)$ and $Y_2 = 2\log(x)$ on your graphing utility. Are they equivalent? What might account for any differences in the two functions?

Enter $Y_1 = \log(x^2)$ into the function editor. Go to `WINDOW` and enter limits for x and y. Since we are not given limits for x or y, use the viewing rectangle $-10 \le x \le 10$ and $-10 \le y \le 10$.

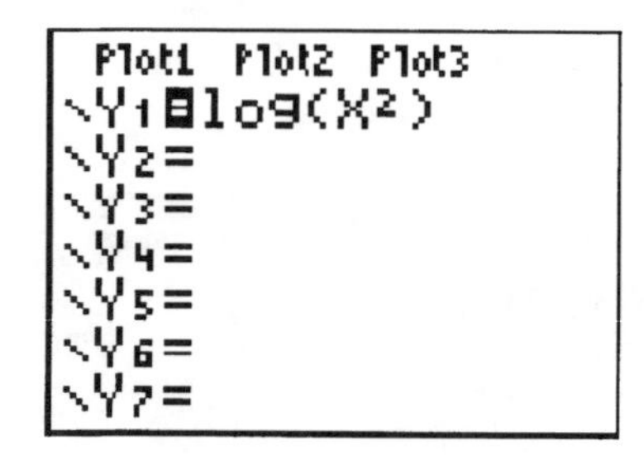

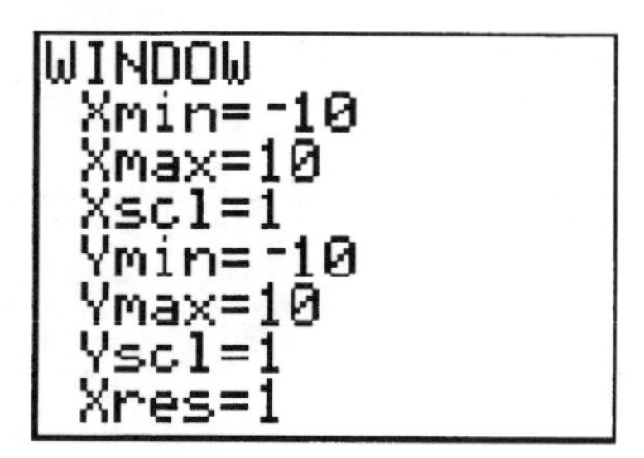

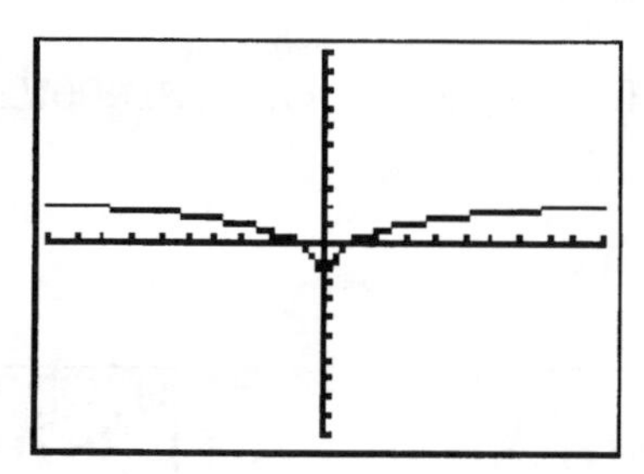

Return to the function editor, clear the first function, enter $Y_2 = 2\log(x)$ and graph on the same view rectangle.

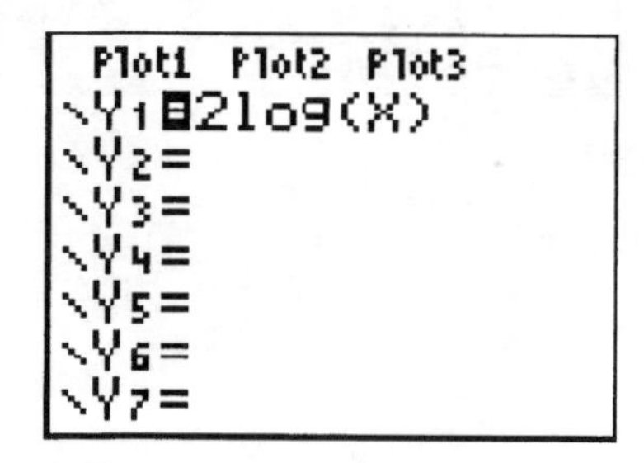

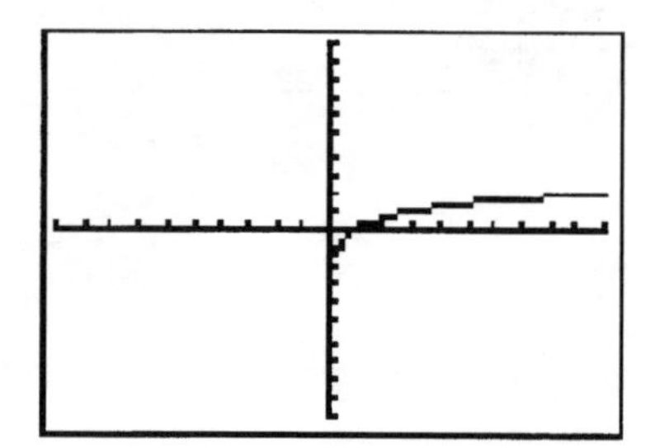

No, they are not equivalent, because the graphs are not identical. The difference is due to the range of x and x^2. The range of x is all real numbers, but the domain of $\log(x)$ is just $x > 0$. The range of x^2 is $x \geq 0$, so the domain of $\log(x^2)$ is all real numbers except 0.

Chapter 11 Review

99. **Minimizing Marginal Cost** The marginal cost of a product can be thought of as the cost of producing one additional unit of output. For example, if the marginal cost of producing the 50th product is \$6.20, then it cost \$6.20 to increase production from 49 to 50 units of output. Callaway Golf Company has determined that the marginal cost C of manufacturing x Big Bertha golf clubs may be expressed by the quadratic function

$$C(x) = 4.9x^2 - 617.4x + 19,600$$

(a) How many golf clubs should be manufactured to minimize the marginal cost?

Enter the formula for C into the function editor. Go to WINDOW and enter limits for x and y. Since we are not given limits for x or y, use the viewing rectangle $0 \leq x \leq 150$ and $0 \leq y \leq 30000$.

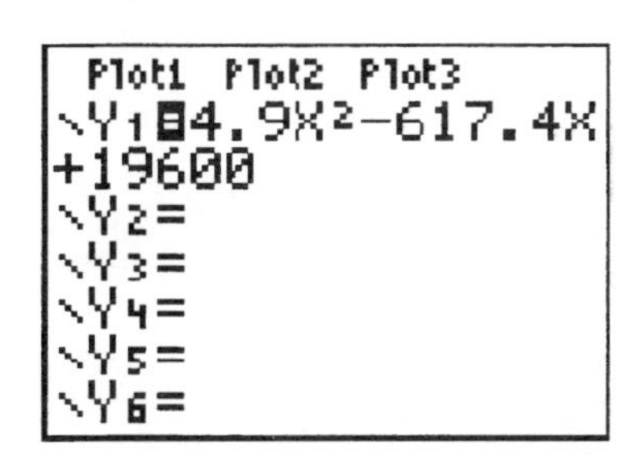

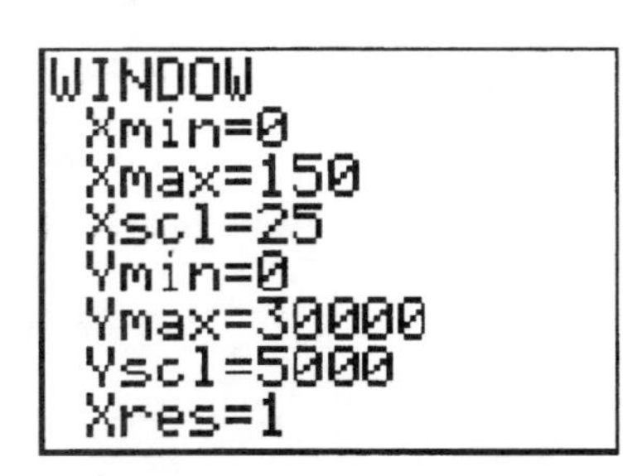

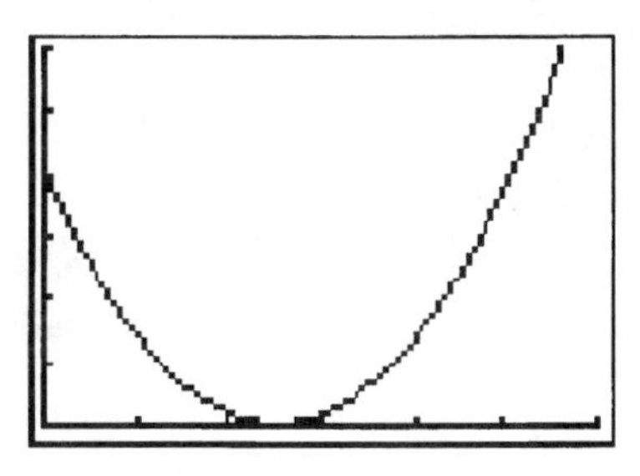

Recall that we can use the `minimum` function to find the local minimum. Note that the x-coordinate of the local minimum is between $x = 50$ and $x = 75$.

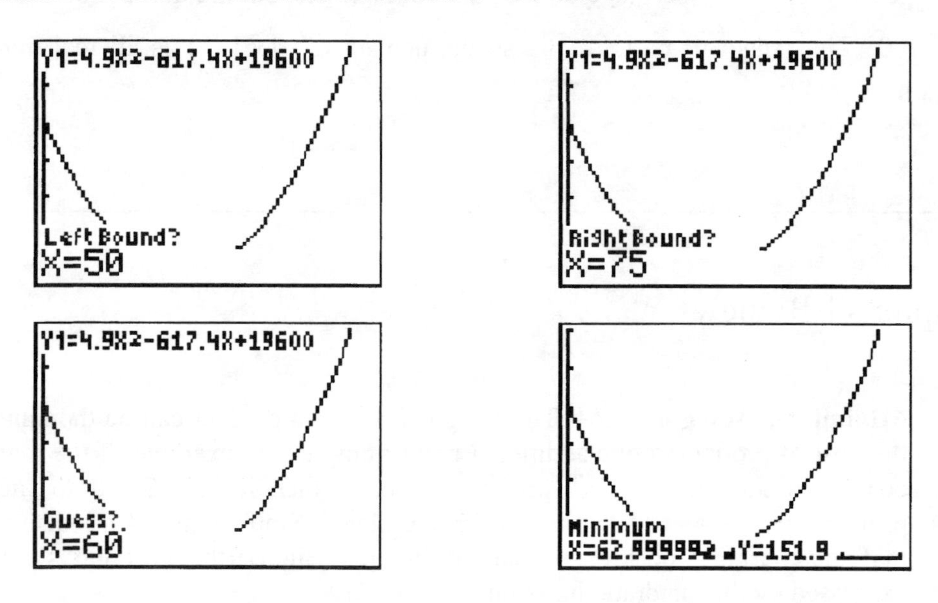

The local minimum is approximately $(63.0, 151.9)$. Thus, Callaway should produce 63 Big Bertha clubs.

(b) At this level of production, what is the marginal cost?

The marginal cost for producing 63 Big Bertha clubs is \$151.90.

Summary

The commands introduced in this chapter were:

```
TRACE

Zoom In
```

Chapter 12 – The Limit of a Function

Section 12.1 Finding Limits Using Tables and Graphs

In Problems 41–46, use a graphing utility to find the indicated limit rounded to two decimal places.

41. $\lim_{x\to 1}\dfrac{x^3-x^2+x-1}{x^4-x^3+2x-2}$

We will determine the limit using a table of values, which we can create using our graphing calculator. There are two options when creating a table using the calculator. We can have the calculator automatically generate a table given a starting value and an increment for x, or we can create a table by picking values for x ourselves. We will use the second option. We must first set the TABLE feature in the correct mode so we can enter values for x.

[2nd] [WINDOW] [▼] [▼] [▶] [ENTER]

Enter the expression into Y_1 in the function editor.

```
Plot1 Plot2 Plot3
\Y1=(X^3-X²+X-1)
/(X^4-X^3+2X-2)
\Y2=
\Y3=
\Y4=
\Y5=
\Y6=
```

Go to TABLE and enter values for x that are close to 1, but less than 1.

.

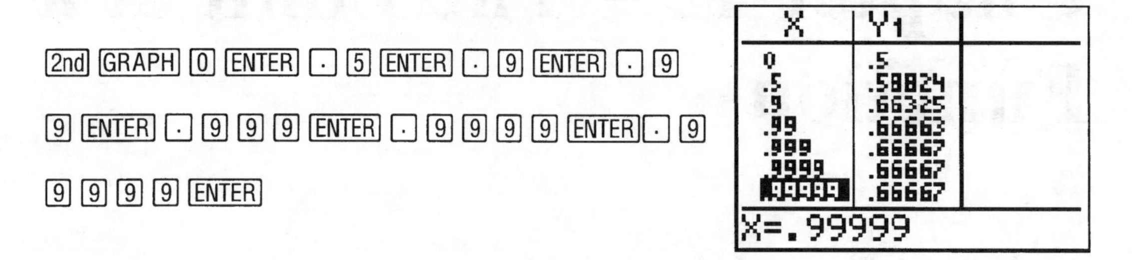

Use the ▲ key to move the cursor to the top of the x column and now enter values for x that are close to 1 but greater than 1.

X	Y1
2	.5
1.5	.60465
1.1	.66346
1.01	.66663
1.001	.66667
1.0001	.66667
1	.66667

X=1.00001

Note that the last entry in the x column appears to be 1. However, since that entry is highlighted, its value is displayed at the bottom of the screen. The entry is really 1.00001. The reason that the calculator displays a 1 in the table is that it only shows entries to five significant digits and 1.00001 rounded to five significant digits is 1.

Thus, rounding to two decimal places,

$$\lim_{x \to 1} \frac{x^3 - x^2 + x - 1}{x^4 - x^3 + 2x - 2} \approx 0.67$$

43. $\lim_{x \to 2} \dfrac{x^3 - 2x^2 + 4x - 8}{x^2 + x - 6}$

Enter the expression into Y_1 in the function editor.

```
Plot1 Plot2 Plot3
\Y1=(X^3-2X²+4X-
8)/(X²+X-6)
\Y2=
\Y3=
\Y4=
\Y5=
\Y6=
```

Go to TABLE and enter values for x that are close to 2, but less than 2.

X	Y1	
1	1.25	
1.5	1.3889	
1.9	1.5531	
1.99	1.5952	
1.999	1.5995	
1.9999	1.6	
2	1.6	

X=1.99999

Note that the last entry in the x column appears to be 2. However, since that entry is highlighted, its value is displayed at the bottom of the screen. The entry is really 1.99999. The reason that the calculator displays a 2 in the table is that it only shows entries to five significant digits and 1.99999 rounded to five significant digits is 2.

Use the ▲ key to move the cursor to the top of the x column and now enter values for x that are close to 2 but greater than 2.

X	Y1	
3	2.1667	
2.5	1.8636	
2.1	1.649	
2.01	1.6048	
2.001	1.6005	
2.0001	1.6	
2	1.6	

X=2.00001

Thus,

$$\lim_{x \to 2} \frac{x^3 - 2x^2 + 4x - 8}{x^2 + x - 6} = 1.6$$

45. $\lim_{x \to -1} \dfrac{x^3 + 2x^2 + x}{x^4 + x^3 + 2x + 2}$

Enter the expression into Y_1 in the function editor.

```
Plot1 Plot2 Plot3
\Y1=(X^3+2X²+X)/
(X^4+X^3+2X+2)
\Y2=
\Y3=
\Y4=
\Y5=
\Y6=
```

Go to TABLE and enter values for x that are close to -1, but less than -1.

X	Y1
-2	-.3333
-1.5	-.5455
-1.1	.16442
-1.01	.01042
-1.001	.001
-1	1E-4
-1	1E-5

X=-1.00001

Note that the last two entries in the x column appears to be -1. However, since the last entry is highlighted, its value is displayed at the bottom. The entry is really -1.00001. The reason that the calculator displays a -1 in the table is that it only shows entries to five significant digits and -1.00001 rounded to five significant digits is -1. The same idea is true for the previous entry.

The outputs 1E-4 and 1E-5 represent numbers in scientific notation. The output 1E-4 represents the number $1\times10^{-4}=0.0001$ and the 1E-5 represents the number $1\times10^{-5}=0.00001$.

Use the ▲ key to move the cursor to the top of the x column and now enter values for x that are close to -1 but greater than -1.

X	Y1
0	0
-.5	-.1333
-.9	-.0708
-.99	-.0096
-.999	-1E-3
-.9999	-1E-4
-1	-1E-5

X=-.99999

The output -1E-4 represents the number $-1\times10^{-4}=-0.0001$ and the -1E-5 represents the number $-1\times10^{-5}=-0.00001$.

Thus,

$$\lim_{x \to -1} \frac{x^3 + 2x^2 + x}{x^4 + x^3 + 2x + 2} = 0$$

Summary

The command introduced in this chapter was:

```
TABLE
```

Chapter 13 – The Derivative of a Function

Section 13.1 The Definition of a Derivative

In Problems 45–54, find the derivative of each function at the given number using a graphing utility.

45. $f(x) = 3x^3 - 6x^2 + 2$ at -2.

We can find the derivative of a function at a point using the nDeriv(function on the graphing calculator. The format of the command is

nDeriv(*function*, *variable*, *value*)

The nDeriv(function is found under the [MATH] menu.

[MATH] [▲] [▲] [▲]

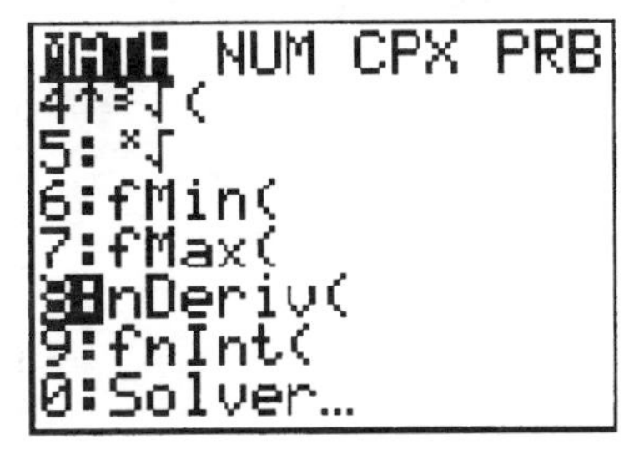

Select the nDeriv(function. Enter the expression for $f(x)$, the name of the variable, and the limit point, separated by commas. Find the value of the derivative.

.

[ENTER] [3] [X,T,Θ,n] [^] [3] [−] [6] [X,T,Θ,n] [x^2] [+] [2] [,]

[X,T,Θ,n] [,] [(−)] [2] [)] [ENTER]

nDeriv(3X^3−6X²+2,X,−2)
60.000003

Thus, rounding the result obtained on the calculator, we have $f'(-2) \approx 60$.

47. $f(x) = \dfrac{-x^3 + 1}{x^2 + 5x + 7}$ at 8

Use nDeriv(to find the value of the derivative at the given point.

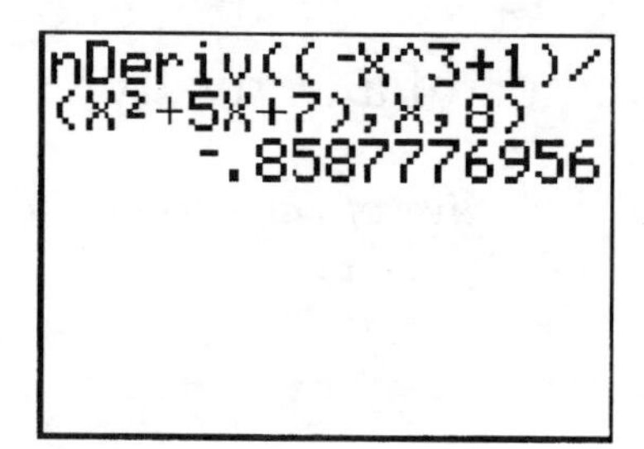

Thus, $f'(8) \approx -0.859$.

49. $f(x) = xe^x$ at 0

Use nDeriv(to find the value of the derivative at the given point.

```
nDeriv(X*e^(X),X
,0)
         1.0000005
```

Thus, rounding the result obtained on the calculator, we have $f'(0) \approx 1$.

51. $f(x) = x^2e^x$ at 1

Use nDeriv(to find the value of the derivative at the given point.

```
nDeriv(X²*e^(X),
X,1)
         8.154851375
```

Thus, $f'(1) \approx 8.155$.

53. $f(x) = xe^{-x}$ at 1

Use `nDeriv(` to find the value of the derivative at the given point.

```
nDeriv(X*e^(-X),
X,1)
          1.22625E-7
```

Recall that the output `1.22625E-7` is the number 1.22625×10^{-7}. Thus, rounding the result obtained on the calculator, we have $f'(1) \approx 0$.

Section 13.5 The Derivatives of the Exponential and Logarithmic Functions; the Chain Rule

73. **Maximizing Profit** At the Super Bowl, the demand for game-day t-shirts is given by

$$p = 50 - 4\ln\left(\frac{x}{100} + 1\right)$$

where p is the price of the shirt in dollars and x is the number of shirts demanded.

(k) Use the TABLE feature of a graphing utility to find the quantity x that maximizes profit.

The profit function is given by $P(x) = 46x - 4x\ln\left(\frac{x}{100}+1\right)$. Enter the formula for the profit function into Y_1 in the function editor. Experiment with a viewing rectangle until your graph shows the maximum.

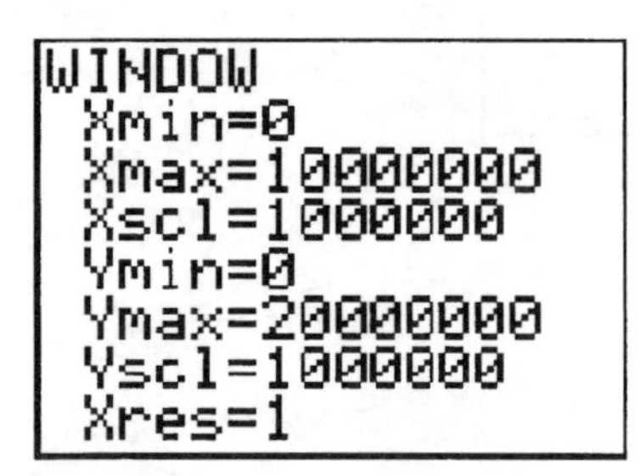

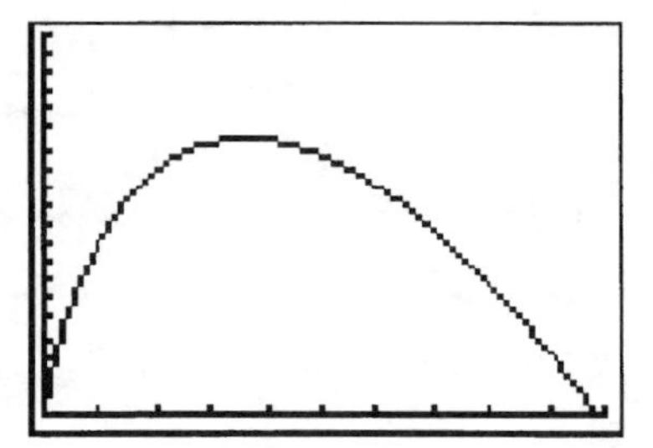

Go to TABLE and enter values for x to find the maximum profit.

X	Y_1
3E6	1.43E7
3.1E6	1.44E7
3.2E6	1.44E7
3.3E6	1.45E7
3.4E6	1.45E7
3.5E6	1.45E7
[illegible]	1.45E7

X=3600000

X	Y_1
3.7E6	1.45E7
3.8E6	1.45E7
3.9E6	1.45E7
4E6	1.45E7
4.1E6	1.44E7
4.2E6	1.44E7
[illegible]	1.43E7

X=4300000

While there appears to be no change in the y-values when x is between 3,300,000 and 4,000,000, the table window is rounding all values to three significant digits. Move the cursor to the second column to investigate the y-values.

X	Y_1
3.3E6	1.45E7
3.4E6	1.45E7
3.5E6	[illegible]
3.6E6	1.45E7
3.7E6	1.45E7
3.8E6	1.45E7
3.9E6	1.45E7

Y_1=14516153.2391

X	Y_1
3.3E6	1.45E7
3.4E6	1.45E7
3.5E6	1.45E7
3.6E6	[illegible]
3.7E6	1.45E7
3.8E6	1.45E7
3.9E6	1.45E7

Y_1=14525251.2745

X	Y_1
3.3E6	1.45E7
3.4E6	1.45E7
3.5E6	1.45E7
3.6E6	1.45E7
3.7E6	[illegible]
3.8E6	1.45E7
3.9E6	1.45E7

Y_1=14523236.7693

Thus, the value of x that maximizes the profit is between 3,500,00 and 3,700,000. Trying values for x in this interval we obtain the following table.

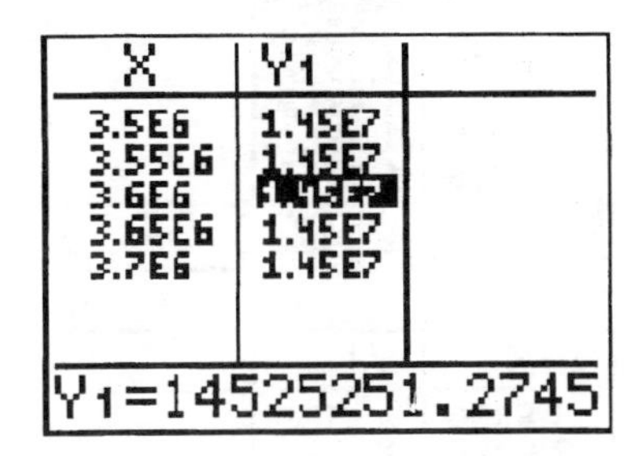

X	Y1
3.5E6	1.45E7
3.55E6	1.45E7
3.6E6	[illegible]
3.65E6	1.45E7
3.7E6	1.45E7

Y1=14525251.2745

X	Y1
3.5E6	1.45E7
3.55E6	1.45E7
3.6E6	1.45E7
3.65E6	[illegible]
3.7E6	1.45E7

Y1=14525613.9278

X	Y1
3.5E6	1.45E7
3.55E6	1.45E7
3.6E6	1.45E7
3.65E6	1.45E7
3.7E6	[illegible]

Y1=14523236.7693

Thus, the value of x that maximizes the profit is between 3,600,00 and 3,700,000. Trying values for x in this interval we obtain the following table.

X	Y1
3.6E6	1.45E7
3.61E6	1.45E7
3.62E6	[illegible]
3.63E6	1.45E7
3.64E6	1.45E7
3.65E6	1.45E7
3.66E6	1.45E7

Y1=14525727.5252

X	Y1
3.6E6	1.45E7
3.61E6	1.45E7
3.62E6	1.45E7
3.63E6	[illegible]
3.64E6	1.45E7
3.65E6	1.45E7
3.66E6	1.45E7

Y1=14525799.7514

X	Y1
3.6E6	1.45E7
3.61E6	1.45E7
3.62E6	1.45E7
3.63E6	1.45E7
3.64E6	[illegible]
3.65E6	1.45E7
3.66E6	1.45E7

Y1=14525761.7847

Thus, the value of x that maximizes the profit is between 3,620,00 and 3,640,000. Trying values for x in this interval we obtain the following table.

X	Y_1
3.63E6	1.45E7
[illegible]	1.45E7
3.63E6	1.45E7
3.63E6	1.45E7
3.63E6	1.45E7
3.64E6	1.45E7
3.64E6	1.45E7

X=3631000

X	Y_1
3.63E6	1.45E7
3.63E6	[illegible]
3.63E6	1.45E7
3.63E6	1.45E7
3.63E6	1.45E7
3.64E6	1.45E7
3.64E6	1.45E7

Y_1=14525800.9085

X	Y_1
3.63E6	1.45E7
3.63E6	1.45E7
[illegible]	1.45E7
3.63E6	1.45E7
3.63E6	1.45E7
3.64E6	1.45E7
3.64E6	1.45E7

X=3632000

X	Y_1
3.63E6	1.45E7
3.63E6	1.45E7
3.63E6	[illegible]
3.63E6	1.45E7
3.63E6	1.45E7
3.64E6	1.45E7
3.64E6	1.45E7

Y_1=14525800.9638

X	Y_1
3.63E6	1.45E7
3.63E6	1.45E7
3.63E6	1.45E7
[illegible]	1.45E7
3.63E6	1.45E7
3.64E6	1.45E7
3.64E6	1.45E7

X=3633000

X	Y_1
3.63E6	1.45E7
3.63E6	1.45E7
3.63E6	1.45E7
3.63E6	[illegible]
3.63E6	1.45E7
3.64E6	1.45E7
3.64E6	1.45E7

Y_1=14525799.9179

Thus, the value of x that maximizes the profit is between 3,631,00 and 3,633,000. We can continue to narrow down the x-interval that contains the maximum, but this is a rather tedious process. A quicker way uses the `maximum` function found in the [MATH] menu.

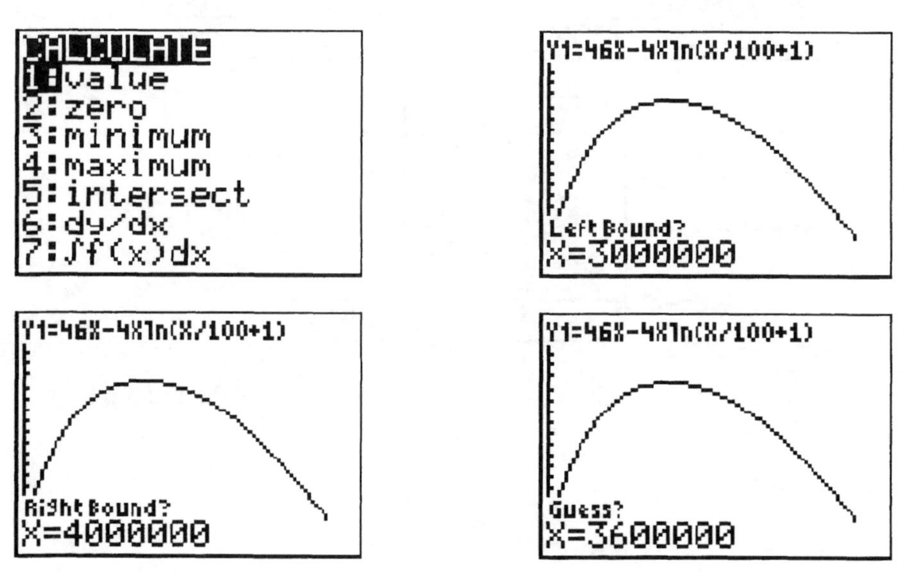

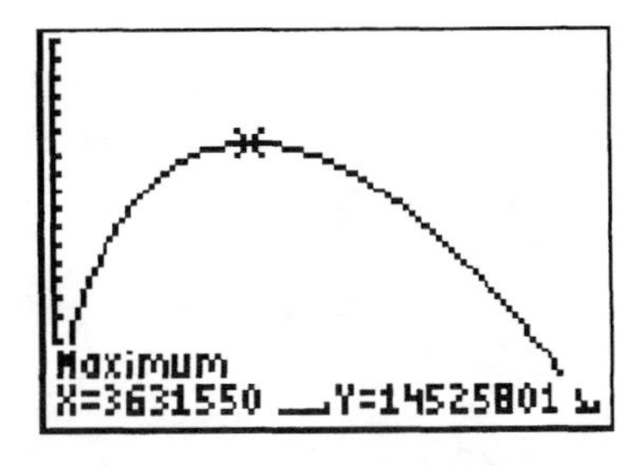

Sell 3,631,550 t-shirts in order to maximize the profit.

(l) What price should be charged for a t-shirt to maximize the profit?

Substitute the value 3,631,550 in for x in the demand equation.

```
50-4ln(3631550/1
00+1)
         7.99989015
```

Rounding to the nearest cent, the price per t-shirt should be $8.00.

Summary

The command introduced in this chapter was:

```
nDeriv(
```

Chapter 14 – Applications: Graphing Functions; Optimization

No Technology Problems in Chapter 14.

Chapter 15 – The Integral of a Function and Applications

Section 15.6 Approximating Definite Integrals

In Problems 17–20, use a graphing utility to approximate each integral. Round your answer to two decimal places.

17. $\int_0^1 e^{x^2}\,dx$

We can approximate the definite integral using the `fnInt(` function on the graphing calculator. The format of the command is

`fnInt(`*function*, *variable*, *lower limit, upper limit*)

The `fnInt(` function is found under the MATH menu.

MATH ▲ ▲

Select the `fnInt(` function. Enter the expression for $f(x)$, the name of the variable, the lower limit, and the upper limit, separated by commas. Find the value of the definite integral.

```
fnInt(e^(X²),X,0
,1)
           1.462651746
```

[ENTER] [2nd] [LN] [X,T,Θ,n] [x²] [)] [,] [X,T,Θ,n] [,] [0] [,] [1]

[)] [ENTER]

Thus, rounding the result obtained on the calculator, we have $\int_0^1 e^{x^2}\,dx \approx 1.46$.

19. $\int_1^5 \frac{e^x}{x}\,dx$

Use `fnInt(` to approximate the value of the definite integral.

```
fnInt(e^(X)/X,X,
1,5)
        38.29015754
```

Thus, $\int_1^5 \frac{e^x}{x}\,dx \approx 38.29$.

21. Consider the function $f(x)=\sqrt{1-x^2}$ whose domain is the closed interval $[-1,1]$.

e. Evaluate the integral using a graphing utility.

The integral that we are asked to approximate is $\int_{-1}^{1}\sqrt{1-x^2}\,dx$. Use `fnInt(` to find approximate the definite integral.

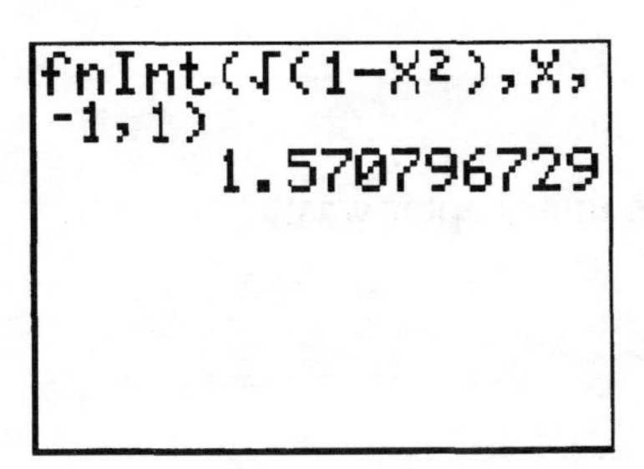

Thus, rounding the result obtained on the calculator, we have $\int_{-1}^{1}\sqrt{1-x^2}\,dx \approx 1.57$. The area under the curve is approximately 1.57 square units.

Summary

The command introduced in this chapter was:

```
fnInt(
```

Chapter 16 – Other Applications and Extensions of the Integral

Section 16.3 Continuous Probability Functions

43. Suppose that X is the length of gestation in healthy humans. Then X is approximately normally distributed with a mean of 280 days and a standard deviation of 10 days. A probability density function for X is given by

$$f(x) = \frac{1}{10\sqrt{2\pi}} e^{-(x-280)^2/200}$$

(a) Use this probability density function to determine the probability that a healthy pregnant woman will have a pregnancy that lasts more than one week beyond the mean for the length of gestation in healthy humans.

We are asked to find $P(X > 287) = \int_{287}^{\infty} \frac{1}{10\sqrt{2\pi}} e^{-(x-280)^2/200} dx$

Use fnInt(to approximate the value of the definite integral.

We do not have ∞ on our calculator, so we cannot use the fnInt(function to evaluate the improper integral directly. Remember that the area under the normal curve is 1 and that the normal curve is symmetric. This implies that

$$\int_{280}^{\infty} \frac{1}{10\sqrt{2\pi}} e^{-(x-280)^2/200} dx = \frac{1}{2}$$

Rewriting the integral on the left side we obtain

$$\int_{280}^{287} \frac{1}{10\sqrt{2\pi}} e^{-(x-280)^2/200} dx + \int_{287}^{\infty} \frac{1}{10\sqrt{2\pi}} e^{-(x-280)^2/200} dx = \frac{1}{2}$$

Solving for the improper integral gives us the following relationship.

$$\int_{287}^{\infty} \frac{1}{10\sqrt{2\pi}} e^{-(x-280)^2/200} dx = \frac{1}{2} - \int_{280}^{287} \frac{1}{10\sqrt{2\pi}} e^{-(x-280)^2/200} dx$$

This equation will allow us to use the `fnInt(` function to approximate the improper integral.

```
1/2-fnInt(1/(10√
(2π))*e^(-(X-280
)²/200),X,280,28
7)
       .2419636522
```

Thus, $P(X > 287) \approx 0.242$.

(b) Determine the probability that a healthy pregnant woman will have a pregnancy such that the length of the pregnancy is within one week of the mean for the length of gestation in healthy humans.

We are asked to find $P(273 < X < 287) = \int_{273}^{287} \frac{1}{10\sqrt{1\pi}} e^{-(x-280)^2/200} dx$

Use `fnInt(` to approximate the value of the definite integral.

```
fnInt(1/(10√(2π)
)*e^(-(X-280)²/2
00),X,273,287)
       .5160726956
```

Thus, $P(273 < X < 287) \approx 0.516$.

Most graphing utilities have a random number function (usually RAND or RND) generating numbers between 0 and 1. Every time you use a random number function, a different number is selected. Check your user's manual to see how to use this feature of your graphing utility.

Sometimes probabilities are found by experimentation, that is, by performing an experiment. In Problems 47 and 48 use a random number function to perform an experiment.

47. Use a random number function to select a value for the random variable X. Repeat this experiment 50 times. [*Note:* Most calculators repeat the action of the last entry if you simply press the ENTER, or EXE, key again.] Count the number of times the random variable X is between 0.6 and 0.9.

    (a) Calculate the ratio

$$R = \frac{\text{Number of times the random variable } X \text{ is between 0.6 and 0.9}}{50}$$

You can generate a list of random numbers using the `rand` command. If you wish to generate n random numbers, the format for the command is

```
rand(n)
```

With 50 entries in the list, it can be difficult to count how many entries are between 0.6 and 0.9. You can have your calculator count number the entries between 0.6 and 0.9 using the `sum` command. The sum command is found in the `MATH` submenu of the `[LIST]` menu. The format for the command is

```
Sum(list)
```

To count the number of values in a list that fall in a desired interval, we test the values in the list using inequalities. To count the values in a list that are between 0.6 and 0.9 we would use

```
Sum(L1>Ø.6 and L1<Ø.9)
```

Remember that an inequality of the form $a < x < b$ is equivalent to $a < x$ **and** $x < b$. The inequality symbols are found under the `[TEST]` menu, and the logical operator `and` is found under the `LOGIC` submenu of `[TEST]`.

Generate a list of 50 random numbers and store the list in L1.

MATH ◄ 1 (5 0) STO► 2nd 1 ENTER

```
rand(50)→L1
{.7114187056 .7...
```

NOTE: We probably won't have the same numbers. The random generator generates random numbers, these will vary from calculator to calculator.

Count the number of random numbers between 0.6 and 0.9.

2nd STAT ◄ 5 2nd 1 2nd MATH 3 0 . 6 2nd

MATH ► 1 2nd 1 2nd MATH 5 0 . 9) ENTER

```
sum(L1>0.6 and L
1<0.9)
                20
```

There were 20 numbers between 0.6 and 0.9. Thus, the ratio is $R = \frac{20}{50} = 0.4$

(b) Calculate the actual probability $Pr(0.6 \le X \le 0.9)$ using a uniform density function.

The uniform density function is $f(x) = \begin{cases} 1 & \text{if } 0 \le x \le 1 \\ 0 & \text{if } x < 0 \text{ or } x > 1 \end{cases}$. The actual probability is given by $Pr(0.6 \le X \le 0.9) = \int_{0.6}^{0.9} f(x)\,dx = \int_{0.6}^{0.9} 1dx$.

```
fnInt(1,X,0.6,0.
9)
                .3
```

Thus, $Pr(0.6 \le X \le 0.9) = 0.3$.

Summary

The new commands introduced in this chapter were:

```
rand(

Sum(

and
```

Chapter 17 – Calculus of Functions of Two or More Variables

No Technology Problems in Chapter 17.

Appendices A - C

Appendix C - Graphing Utilities

Appendix C.1 The Viewing Rectangle

In Problems 1–4, determine the coordinates of the points shown. Tell in which quadrant each point lies. Assume that the coordinates are integers.

1.

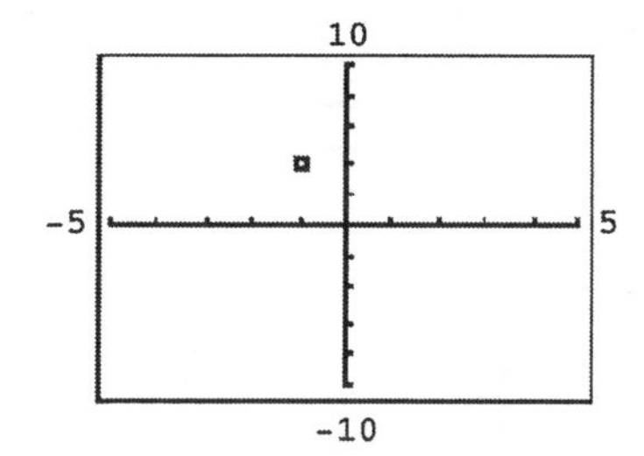

The point is $(-1,4)$, which lies in Quadrant II.

3.

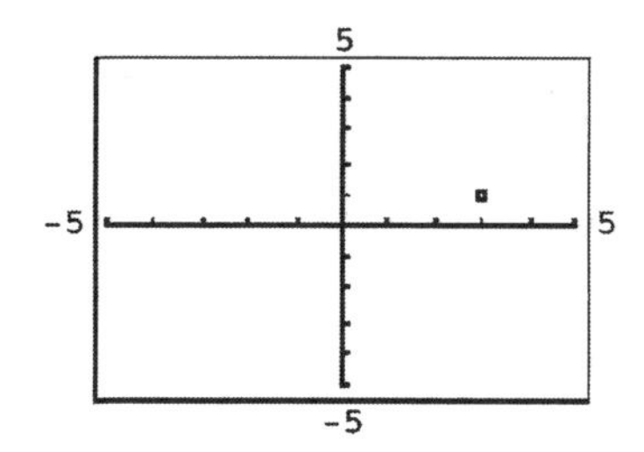

The point is $(3,1)$, which lies in Quadrant I.

In Problems 5–10, determine the viewing window used.

5.

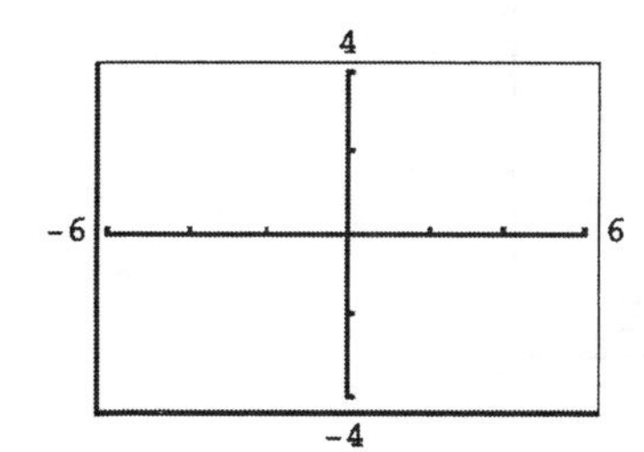

The viewing window used is

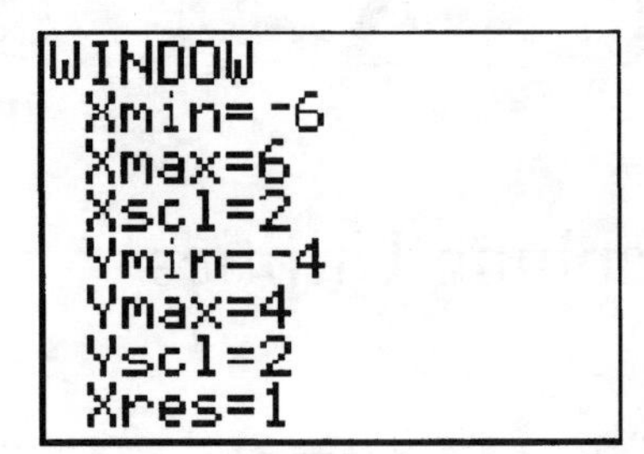

7.

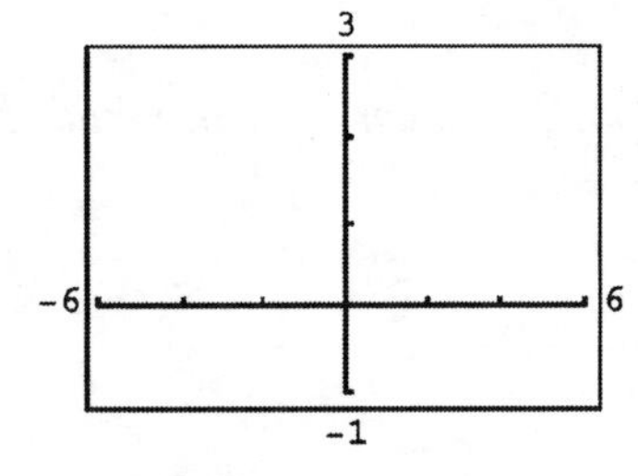

The viewing window used is

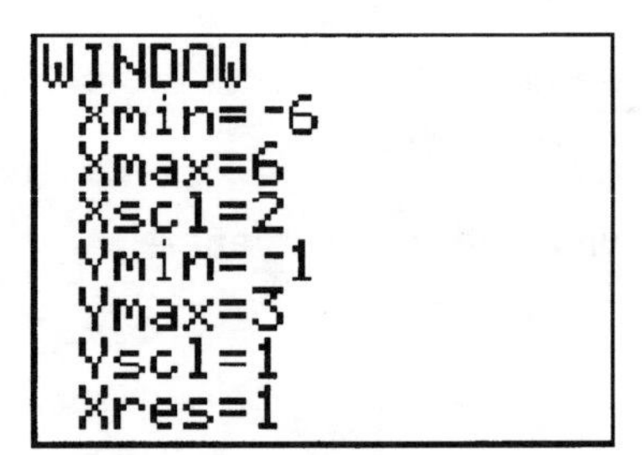

9.

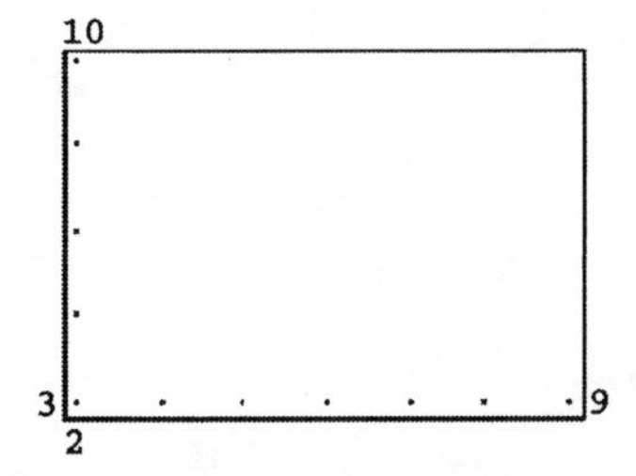

The viewing window used is

WINDOW
Xmin=3
Xmax=9
Xscl=1
Ymin=2
Ymax=10
Yscl=2
Xres=1

In Problems 11–16, select a setting so that each of the given points will lie within the viewing rectangle.

11. $(-10,5)$, $(3,-2)$, $(4,-1)$

Answers will vary. One possible viewing rectangle is

```
WINDOW
  Xmin=-12
  Xmax=6
  Xscl=1
  Ymin=-4
  Ymax=7
  Yscl=1
  Xres=1
```

13. $(40,20)$, $(-20,-80)$, $(10,40)$

Answers will vary. One possible viewing rectangle is

```
WINDOW
  Xmin=-30
  Xmax=50
  Xscl=10
  Ymin=-100
  Ymax=50
  Yscl=10
  Xres=1
```

15. $(0,0)$, $(100,5)$, $(5,150)$

Answers will vary. One possible viewing rectangle is

```
WINDOW
  Xmin=-10
  Xmax=110
  Xscl=10
  Ymin=-20
  Ymax=180
  Yscl=20
  Xres=1
```

Appendix C.2 Using a Graphing Utility to Graph Equations

In Problems 1–16, graph each equation using the following windows:

(a)	(b)	(c)	(d)
Xmin = −5	Xmin = −10	Xmin = −10	Xmin = −5
Xmax = 5	Xmax = 10	Xmax = 10	Xmax = 5
Xscl = 1	Xscl = 1	Xscl = 2	Xscl = 1
Ymin = −4	Ymin = −8	Ymin = −8	Ymin = −20
Ymax = 4	Ymax = 8	Ymax = 8	Ymax = 20
Yscl = 1	Yscl = 1	Yscl = 2	Yscl = 5

1. $y = x + 2$

a.

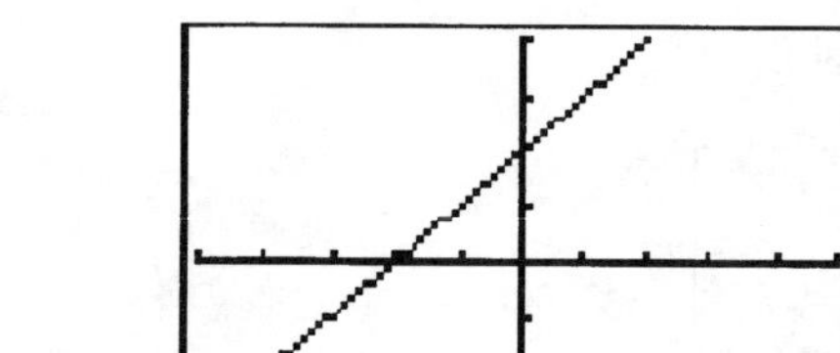

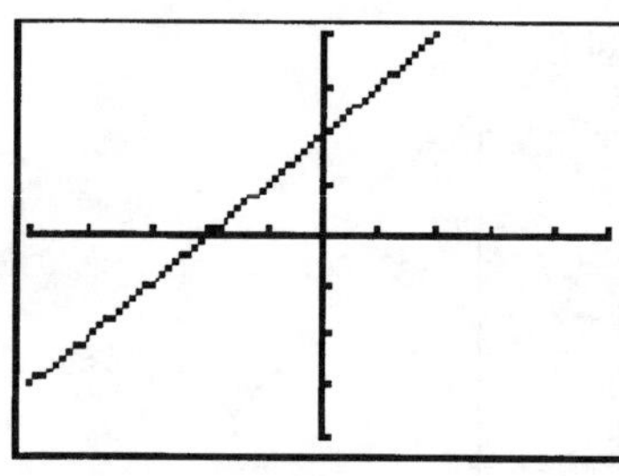

b.

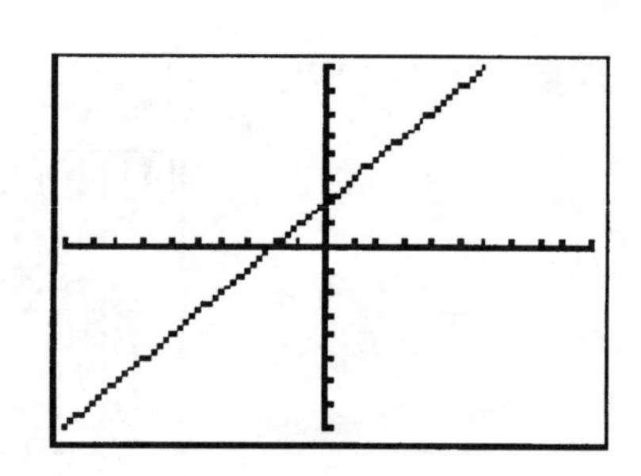

c.

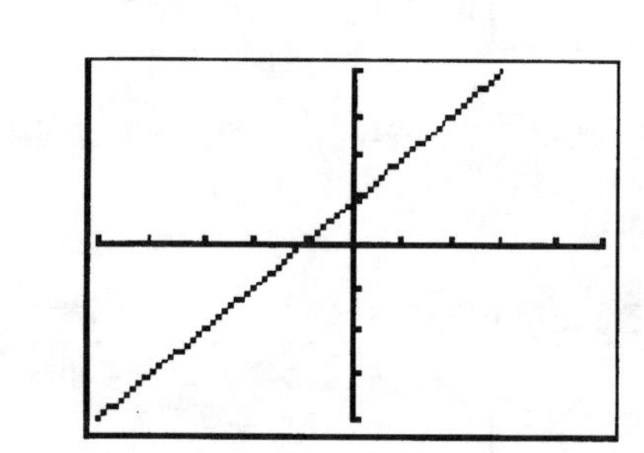

d.

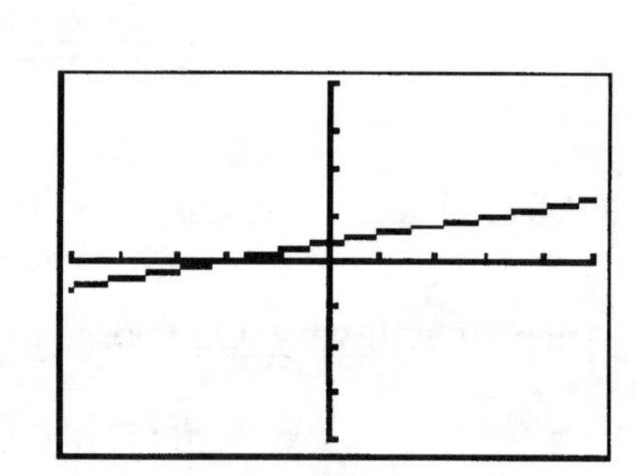

3. $y = -x + 2$

a.

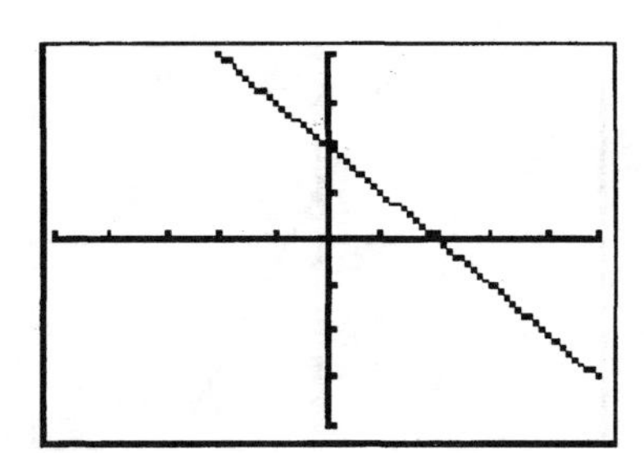

b.

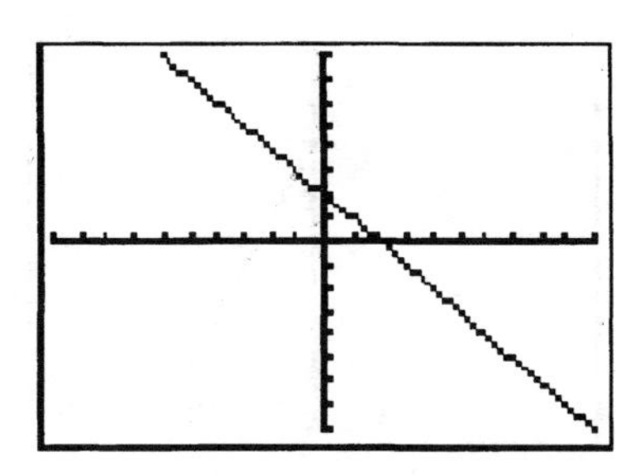

c.

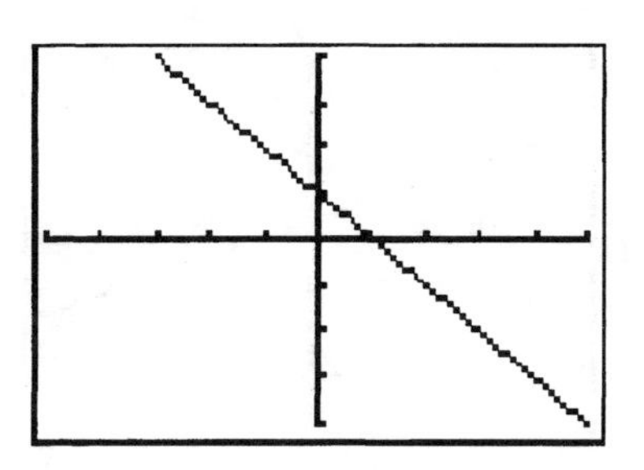

d.

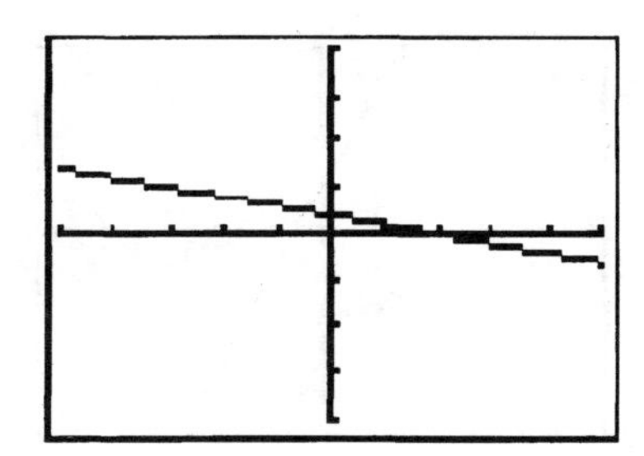

5. $y = 2x + 2$

a.

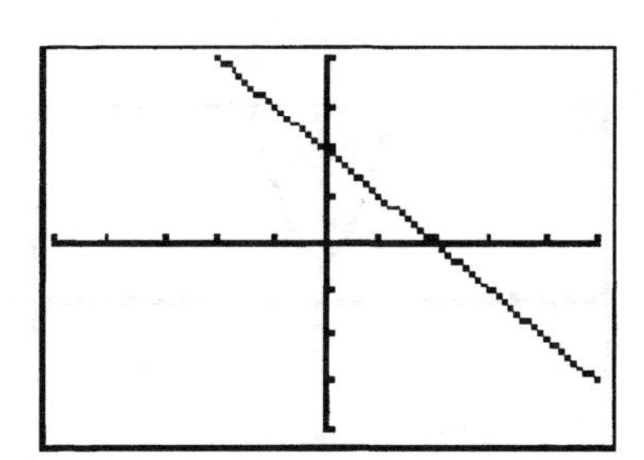

b.

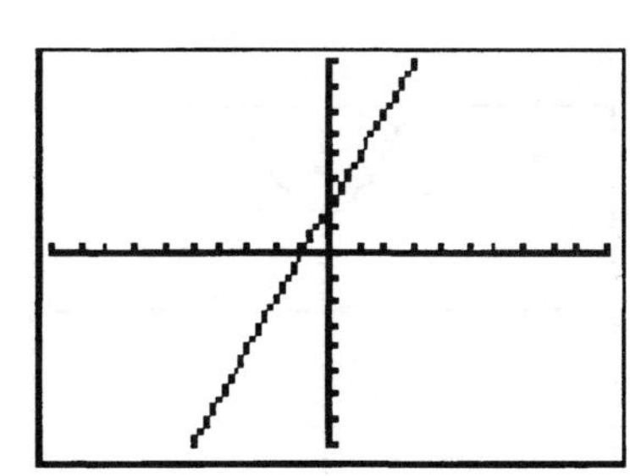

c.

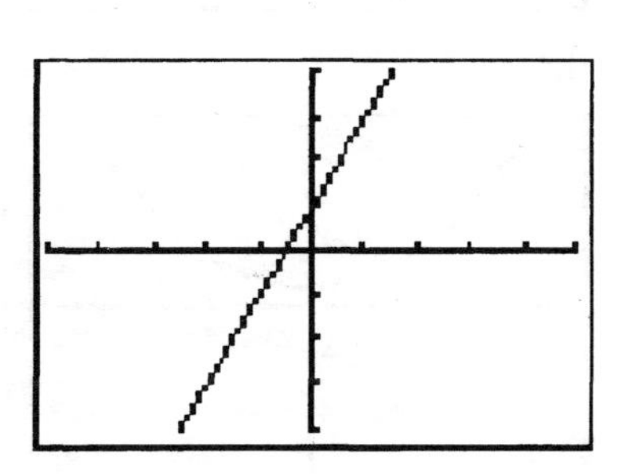

d.

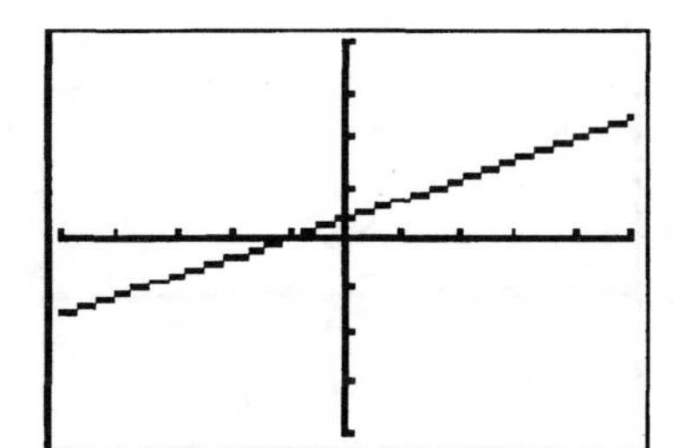

7. $y = -2x + 2$

a.

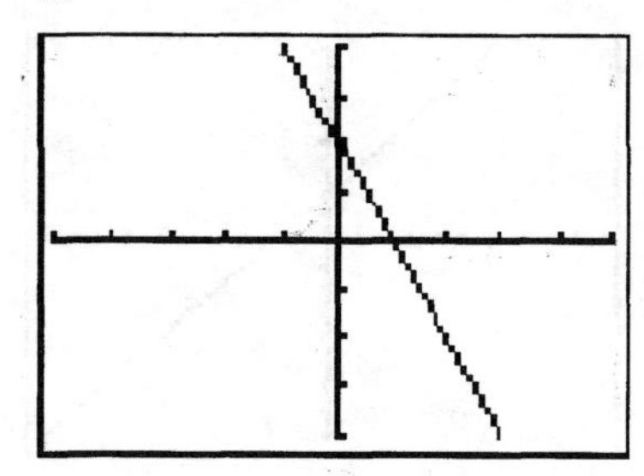

b.

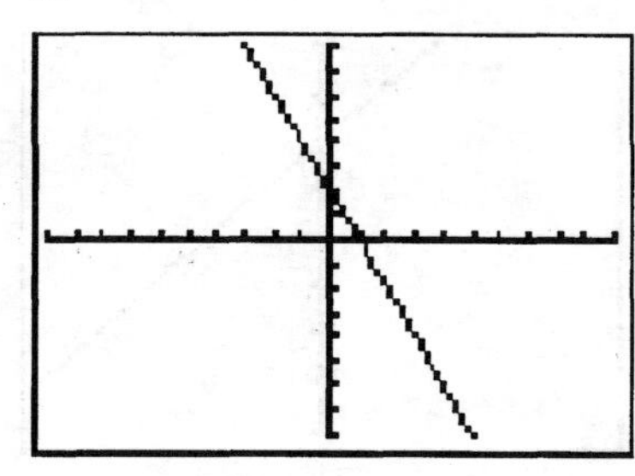

c.

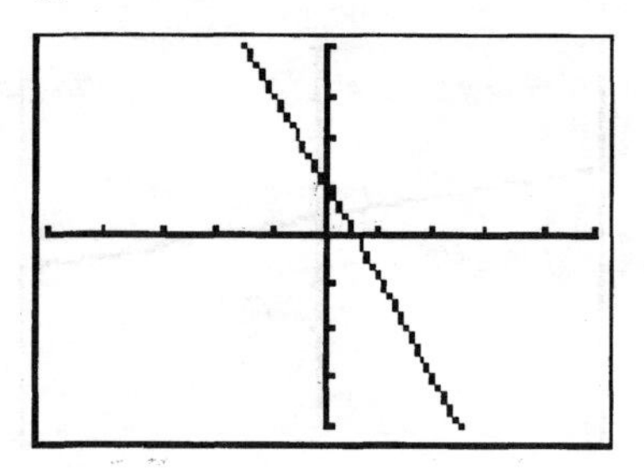

d.

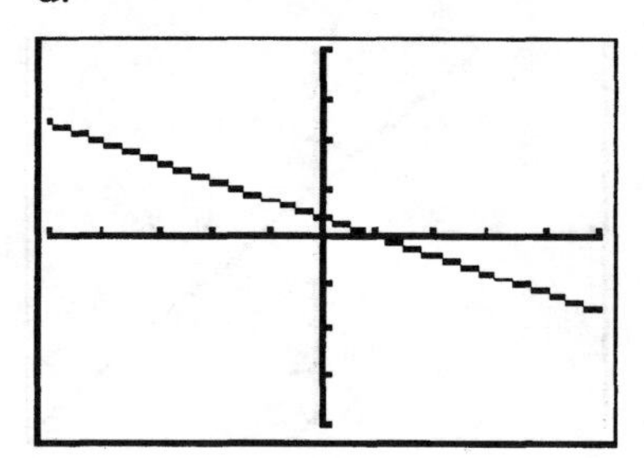

9. $y = x^2 + 2$

a.

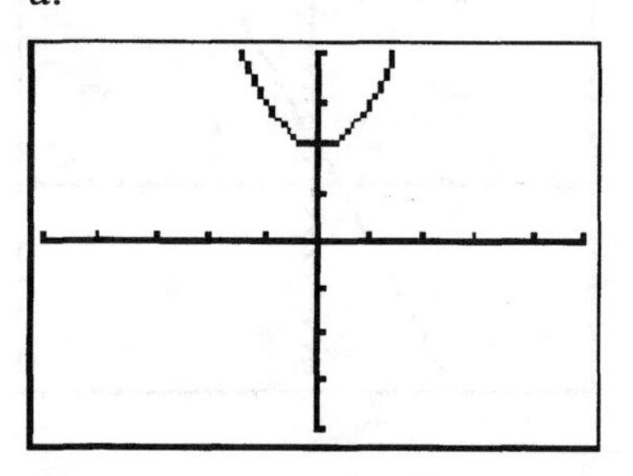

b.

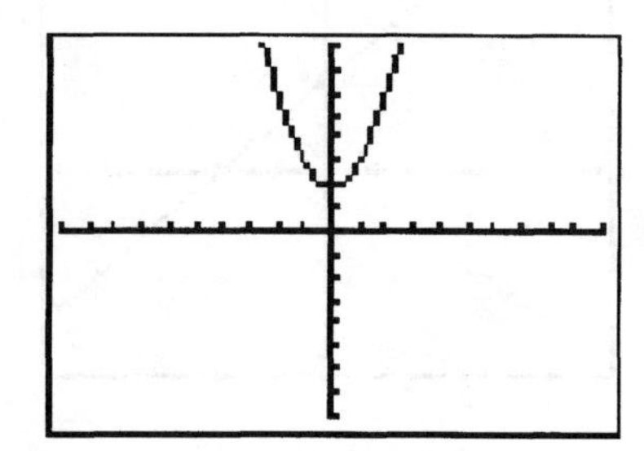

c.

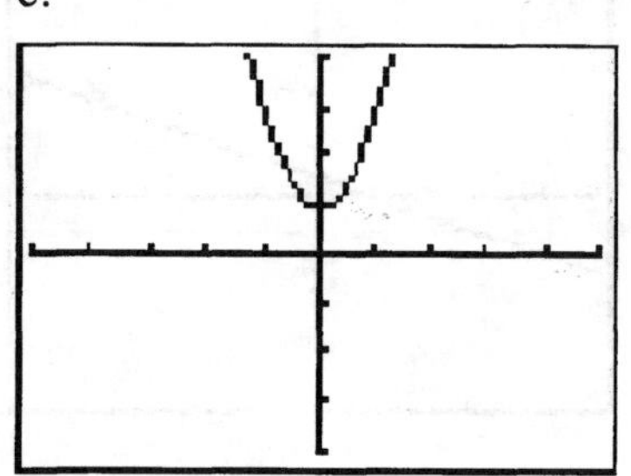

d.

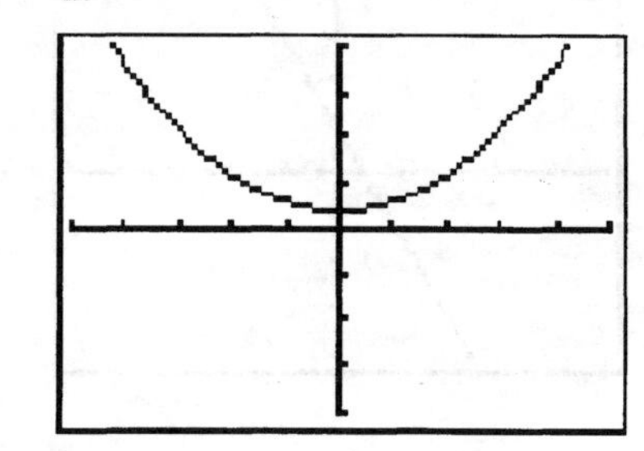

11. $y = -x^2 + 2$

a.

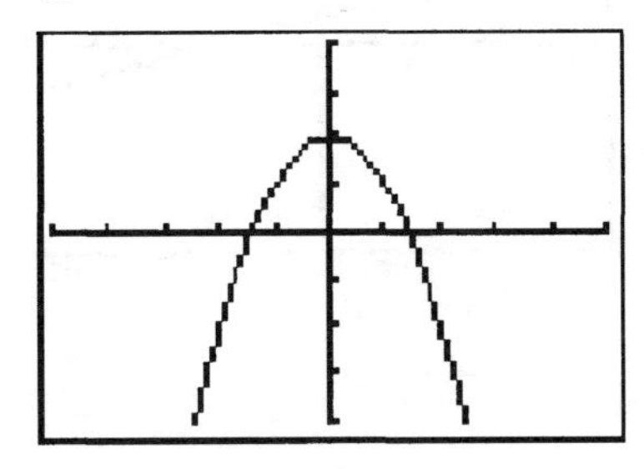

b.

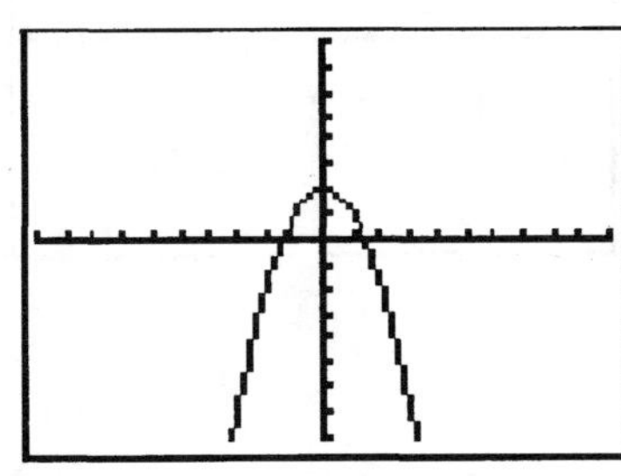

c.

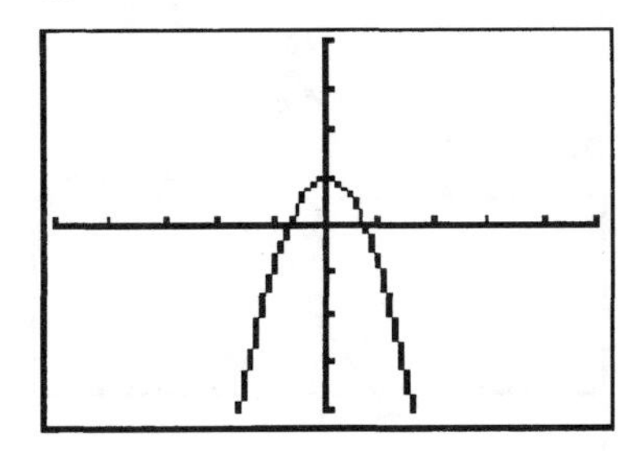

d.

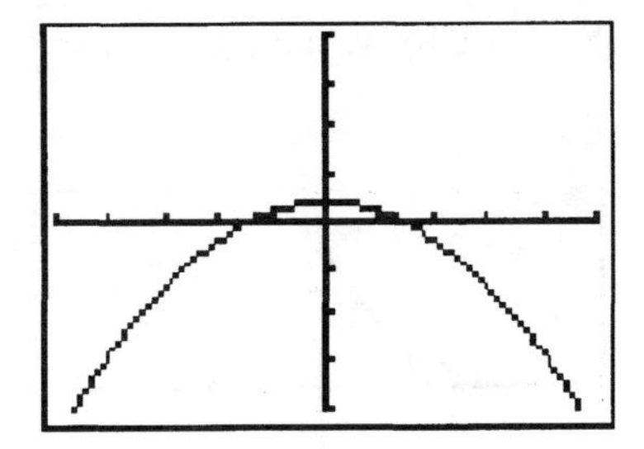

13. $3x + 2y = 6$

Solving for y, we obtain $y = -\frac{3}{2}x + 3$.

a.

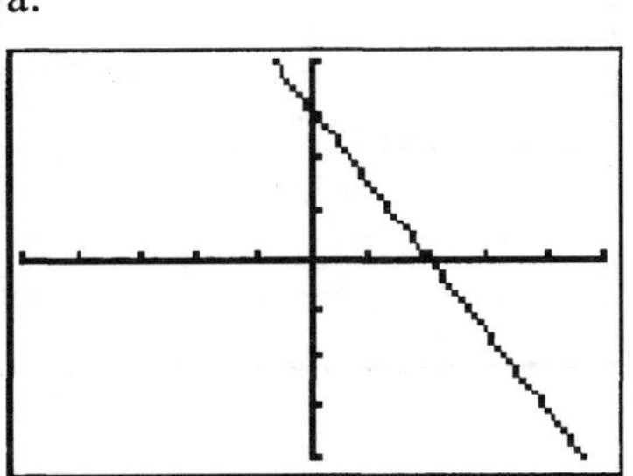

b.

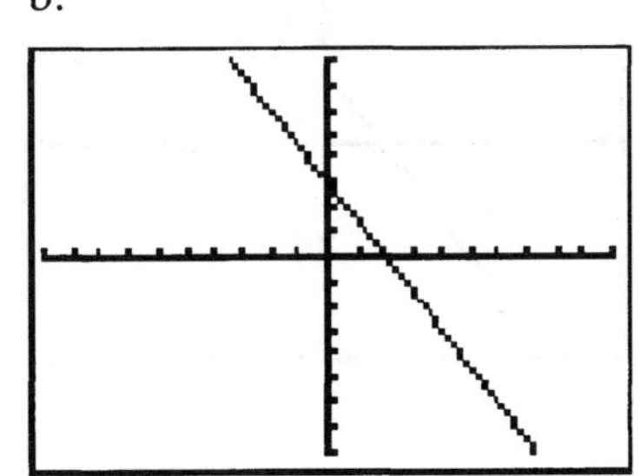

c.

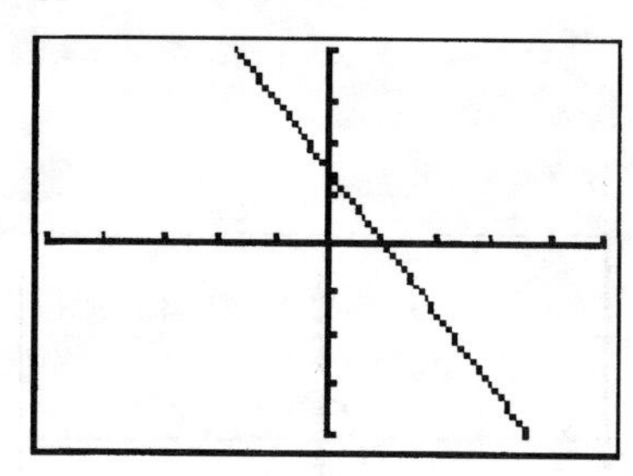

d.

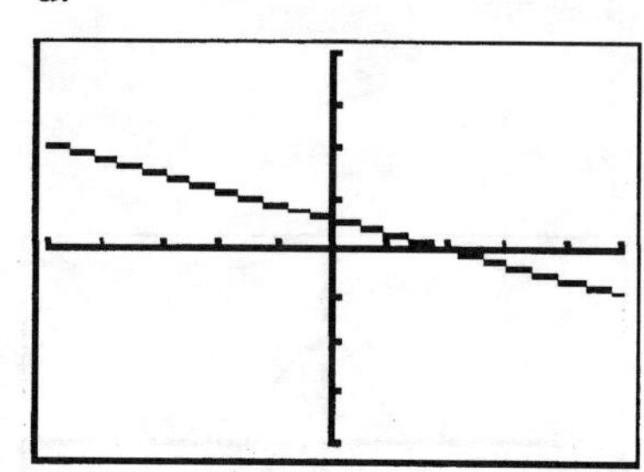

15. $-3x+2y=6$

Solving for y, we obtain $y=\frac{3}{2}x+3$.

a.

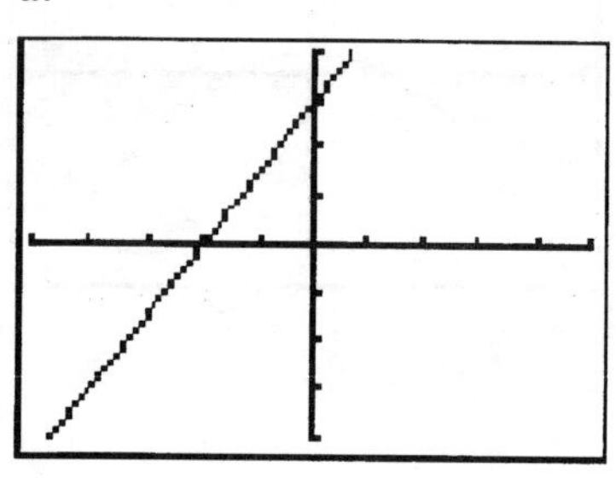

b.

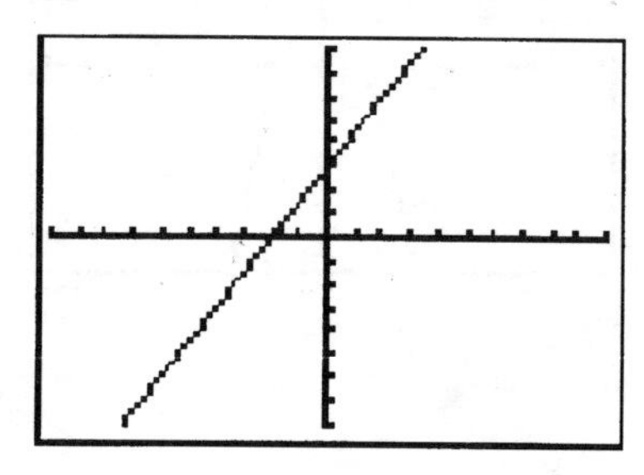

c.

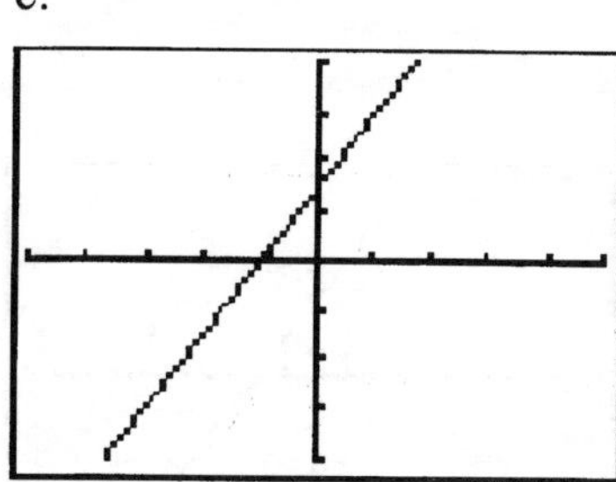

d.

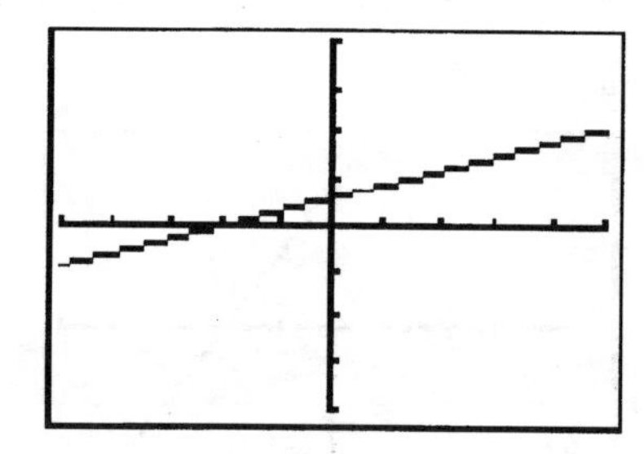

Problems 17–32. For each of the above equations, create a table, $-3 \le x \le 3$, *and list the points on the graph.*

17. $y=x+2$

Go to [Y=] and enter the equation in the function editor.

Plot1 Plot2 Plot3
\Y1 = X+2
\Y2 =
\Y3 =
\Y4 =
\Y5 =
\Y6 =
\Y7 =

To create a table of values we will use the [TABLE] feature on the calculator. First, set up the calculator so it will generate a table using integer values from –3 to 3 for *x*. We can do this in the [TBLSET] menu.

[2nd] [WINDOW] [(-)] [3] [▼] [1] [▼] [ENTER] [▼] [ENTER]

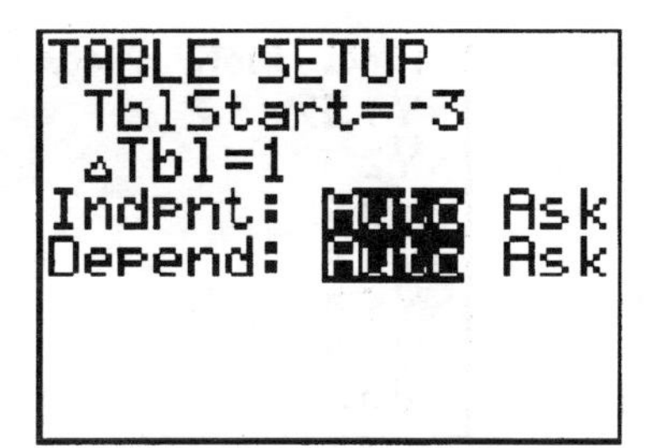

Create the table of values, where $-3 \le x \le 3$.

[2nd] [GRAPH]

X	Y1
-3	-1
-2	0
-1	1
0	2
1	3
2	4
3	5

X=3

19. $y = -x + 2$

Go to [Y=] and enter the equation in the function editor. Create the table of values, where $-3 \le x \le 3$.

Plot1 Plot2 Plot3
\Y1 = -X+2
\Y2 =
\Y3 =
\Y4 =
\Y5 =
\Y6 =
\Y7 =

X	Y1
-3	5
-2	4
-1	3
0	2
1	1
2	0
3	-1

X=3

21. $y = 2x + 2$

Go to [Y=] and enter the equation in the function editor. Create the table of values, where $-3 \le x \le 3$.

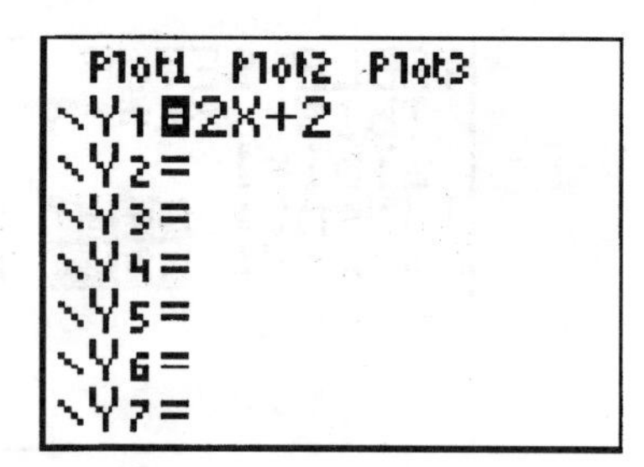

X	Y1
-3	-4
-2	-2
-1	0
0	2
1	4
2	6
3	8

X=3

23. $y = -2x + 2$

Go to [Y=] and enter the equation in the function editor. Create the table of values, where $-3 \le x \le 3$.

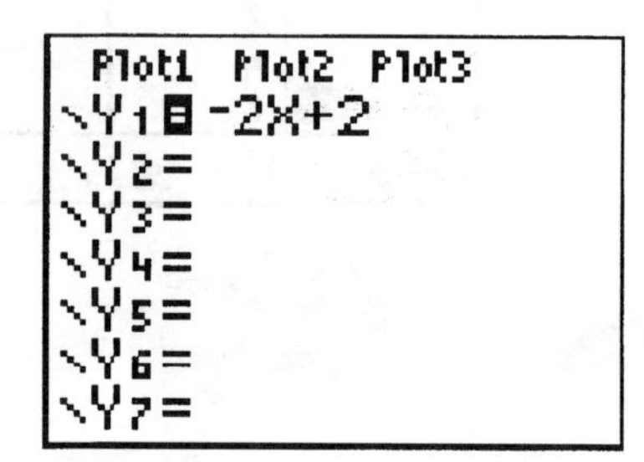

X	Y1
-3	8
-2	6
-1	4
0	2
1	0
2	-2
3	-4

X=3

25. $y = x^2 + 2$

Go to [Y=] and enter the equation in the function editor. Create the table of values, where $-3 \le x \le 3$.

Plot1 Plot2 Plot3
\Y1=X²+2
\Y2=
\Y3=
\Y4=
\Y5=
\Y6=
\Y7=

X	Y1
-3	11
-2	6
-1	3
0	2
1	3
2	6
3	11

X=3

27. $y=-x^2+2$

Go to [Y=] and enter the equation in the function editor. Create the table of values, where $-3 \le x \le 3$.

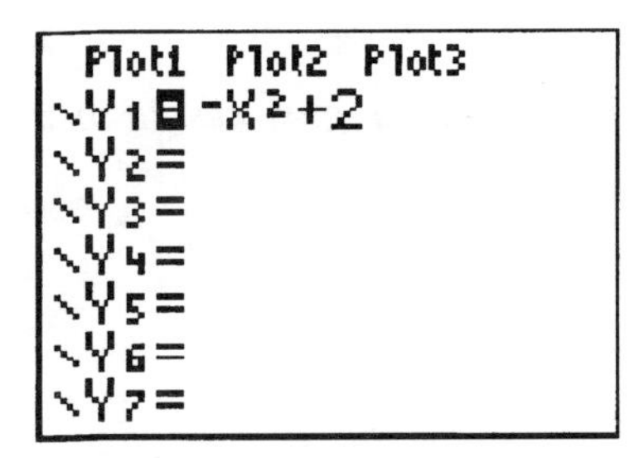

X	Y1
-3	-7
-2	-2
-1	1
0	2
1	1
2	-2
3	-7

X=3

29. $3x+2y=6$

Solve the equation for *y*. In this case, we obtain $y=-\frac{3}{2}x+3$. Go to [Y=] and enter the equation in the function editor. Create the table of values, where $-3 \le x \le 3$.

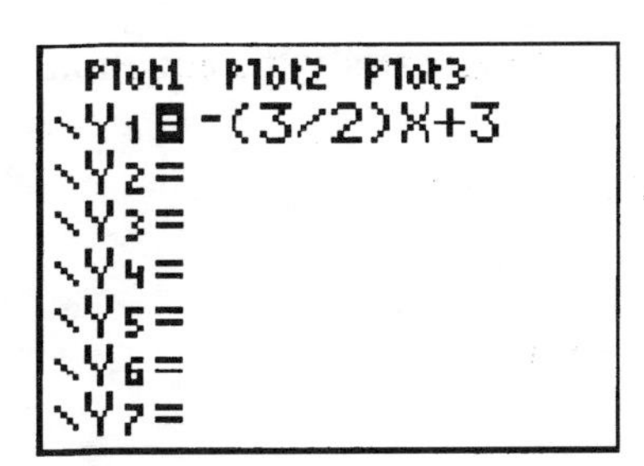

X	Y1
-3	7.5
-2	6
-1	4.5
0	3
1	1.5
2	0
3	-1.5

X=3

31. $-3x+2y=6$

Solve the equation for *y*. In this case, we obtain $y=\frac{3}{2}x+3$. Go to [Y=] and enter the equation in the function editor. Create the table of values, where $-3 \le x \le 3$.

Plot1 Plot2 Plot3
\Y1 = (3/2)X+3
\Y2 =
\Y3 =
\Y4 =
\Y5 =
\Y6 =
\Y7 =

X	Y1
-3	-1.5
-2	0
-1	1.5
0	3
1	4.5
2	6
3	7.5

X=3

Appendix C.3 Square Screens

In Problems 1–8, determine which of the given viewing rectangles result in a square screen.

1.

$$\begin{aligned} \text{Xmin} &= -3 \\ \text{Xmax} &= 3 \\ \text{Xscl} &= 2 \\ \text{Ymin} &= -2 \\ \text{Ymax} &= 2 \\ \text{Yscl} &= 2 \end{aligned}$$

Check to see if the values for Xmin, Xmax, Ymin, and Ymax satisfy the equation

$$2(\text{Xmax} - \text{Xmin}) = 3(\text{Ymax} - \text{Ymin}).$$

$$2(3-(-3)) \stackrel{?}{=} 3(2-(-2))$$
$$12 = 12$$

Yes, this viewing rectangle results in a square window.

3.

$$\begin{aligned} \text{Xmin} &= 0 \\ \text{Xmax} &= 9 \\ \text{Xscl} &= 3 \\ \text{Ymin} &= -2 \\ \text{Ymax} &= 4 \\ \text{Yscl} &= 2 \end{aligned}$$

Check to see if the values for Xmin, Xmax, Ymin, and Ymax satisfy the equation

$$2(\text{Xmax} - \text{Xmin}) = 3(\text{Ymax} - \text{Ymin}).$$

$$\begin{gathered} 2(9-0) \stackrel{?}{=} 3(4-(-2)) \\ 18 = 18 \end{gathered}$$

Yes, this viewing rectangle results in a square window.

5.

$$\begin{aligned} \text{Xmin} &= -6 \\ \text{Xmax} &= 6 \\ \text{Xscl} &= 1 \\ \text{Ymin} &= -2 \\ \text{Ymax} &= 2 \\ \text{Yscl} &= 0.5 \end{aligned}$$

Check to see if the values for Xmin, Xmax, Ymin, and Ymax satisfy the equation

$$2(\text{Xmax} - \text{Xmin}) = 3(\text{Ymax} - \text{Ymin}).$$

$$\begin{gathered} 2(6-(-6)) \stackrel{?}{=} 3(2-(-2)) \\ 24 \neq 12 \end{gathered}$$

No, this viewing rectangle does not result in a square window.

7.

$$\begin{aligned} \text{Xmin} &= 0 \\ \text{Xmax} &= 9 \\ \text{Xscl} &= 1 \\ \text{Ymin} &= -2 \\ \text{Ymax} &= 4 \\ \text{Yscl} &= 1 \end{aligned}$$

Check to see if the values for Xmin, Xmax, Ymin, and Ymax satisfy the equation

$$2(\text{Xmax} - \text{Xmin}) = 3(\text{Ymax} - \text{Ymin}).$$

$$2(9-0) \stackrel{?}{=} 3(4-(-2))$$
$$18 = 18$$

Yes, this viewing rectangle results in a square window.

9. If $\text{Xmin} = -4$, $\text{Xmax} = 8$, and $\text{Xscl} = 1$, how should Ymin, Ymax, and Yscl be selected so that the viewing rectangle contains the point $(4,8)$ and the screen is square?

Given $\text{Xmin} = -4$ and $\text{Xmax} = 8$, we must find values for Ymin, and Ymax that satisfy the equation

$$2(\text{Xmax} - \text{Xmin}) = 3(\text{Ymax} - \text{Ymin}).$$

$$\begin{aligned} 2(8-(-4)) &= 3(\text{Ymax} - \text{Ymin}) \\ 24 &= 3(\text{Ymax} - \text{Ymin}) \\ 8 &= \text{Ymax} - \text{Ymin} \end{aligned}$$

Since the point $(4,8)$ must be in the screen, one possible choice is $\text{Ymin} = 1$ and $\text{Ymax} = 9$ with $\text{Yscl} = 1$. Then,

$$8 \stackrel{?}{=} \text{Ymax} - \text{Ymin}$$
$$8 = 9 - 1 = 8$$

Appendix C.5 Using a Graphing Utility to Locate Intercepts and Check for Symmetry

In Problems 1–6, use ZERO (or ROOT) to approximate the smaller of the two x-intercepts of each equation. Express the answer rounded to two decimal places.

1. $y = x^2 + 4x + 2$

You can use your TI-83 Plus to find the x-intercepts using the `zero` function. The `zero` function will find the zeros (an x-intercept is a zero) of a function. The `zero` finder requires three inputs, an x-value to the left of the zero (x-intercept), an x-value to the right of the zero (x-intercept), and an estimate of the x-value of the zero (x-intercept). The `zero` function is found under the `[CALC]` menu.

First, enter the equation in the function editor. Then, use a standard window and graph.

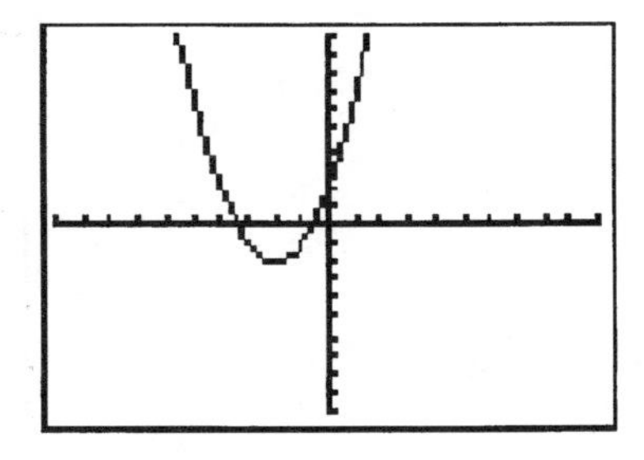

Find the smaller x-intercept. Enter a value for x that is less than (to the left of) the x–intercept. Notice that the smaller x-intercept is between $x = -4$ and $x = -3$, so we can use $x = -4$ as a left bound.

[2nd] [TRACE] [2] [(-)] [4]

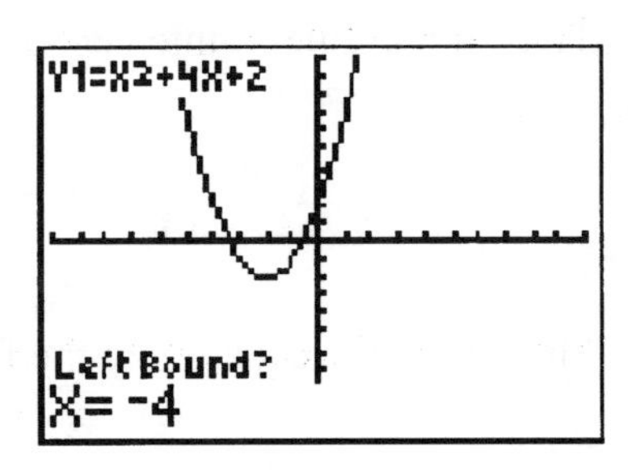

Input $x = -3$ as a right bound.

[ENTER] [(-)] [3]

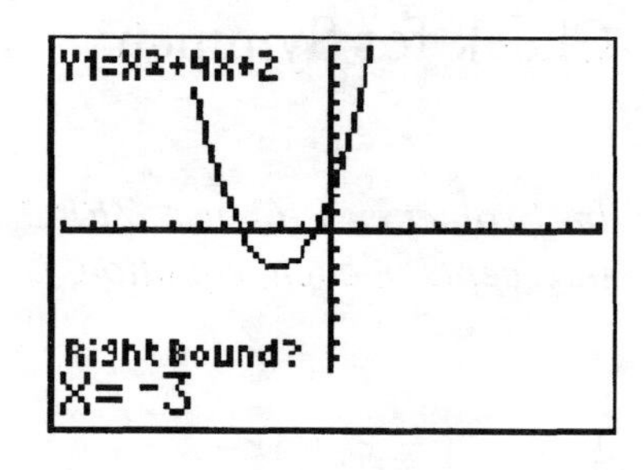

Input $x = -3.5$ as a guess.

[ENTER] [(-)] [3] [.] [5]

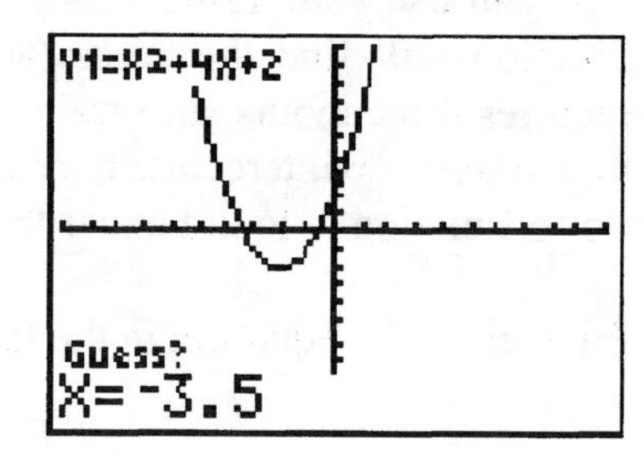

Find the x-intercept.

[ENTER]

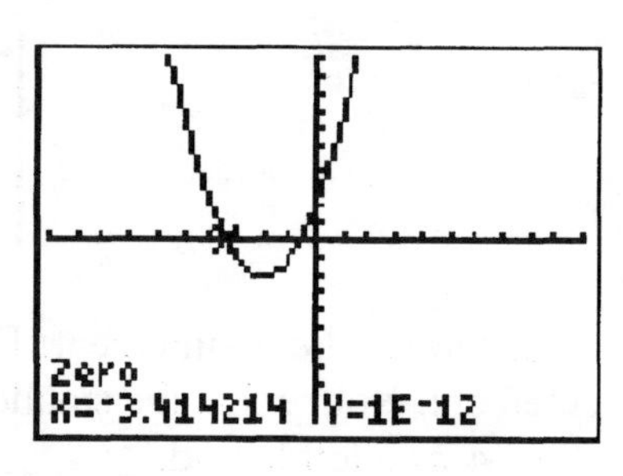

Thus, the smaller x-intercept is approximately $(-3.41, 0)$.

3. $y = 2x^2 + 4x + 1$

First, enter the equation in the function editor. Then, use a standard window and graph.

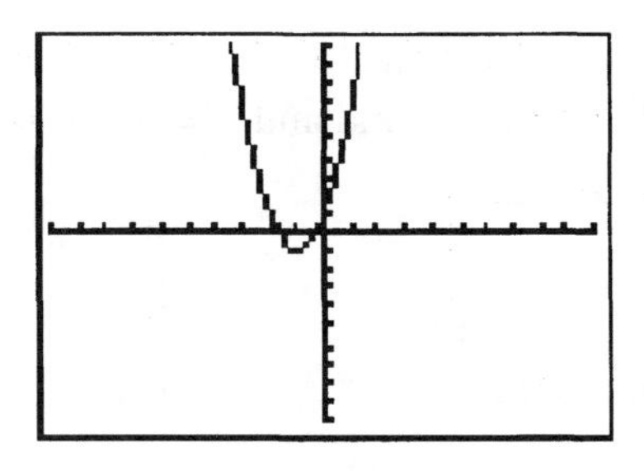

Find the smaller x-intercept. Notice that the smaller x-intercept is between $x=-3$ and $x=-1$, so we can use $x=-3$ as a left bound and $x=-1$ as a right bound.

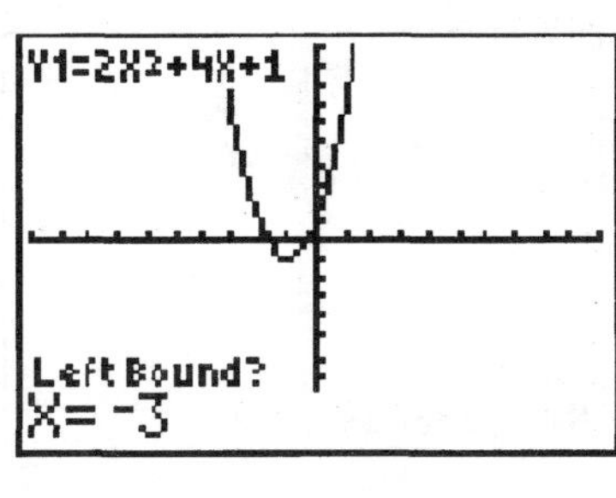

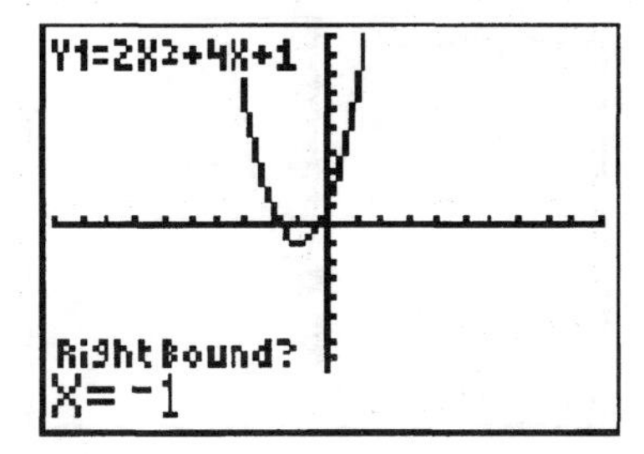

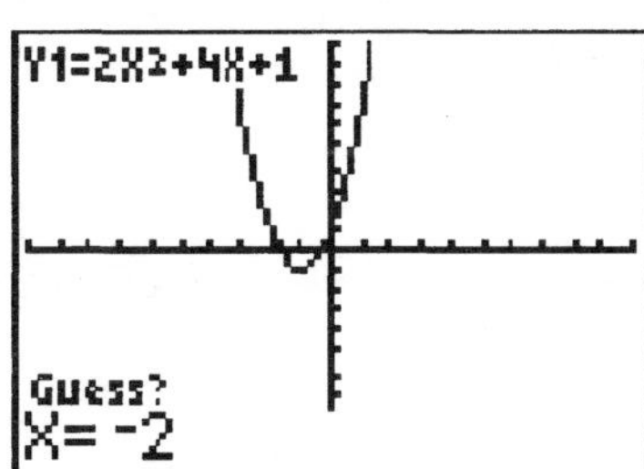

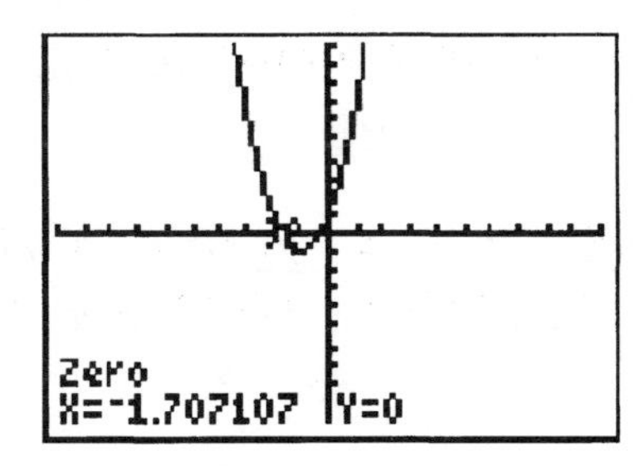

Thus, the smaller x-intercept is approximately $(-1.71, 0)$.

5. $y=2x^2-3x-1$

First, enter the equation in the function editor. Then, use a standard window and graph.

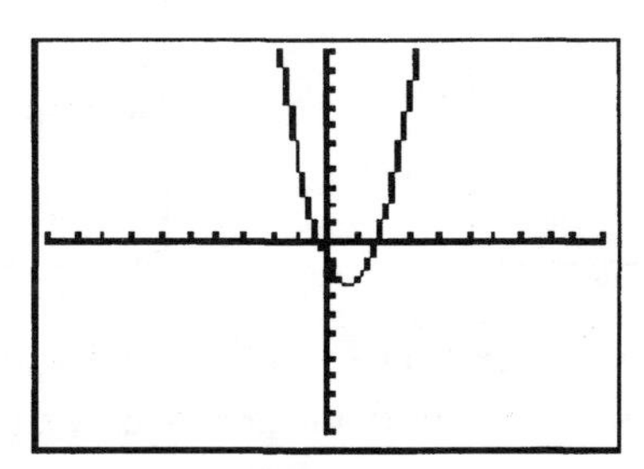

Find the smaller x-intercept. Notice that the smaller x-intercept is between $x=-1$ and $x=0$, so we can use $x=-1$ as a left bound and $x=0$ as a right bound.

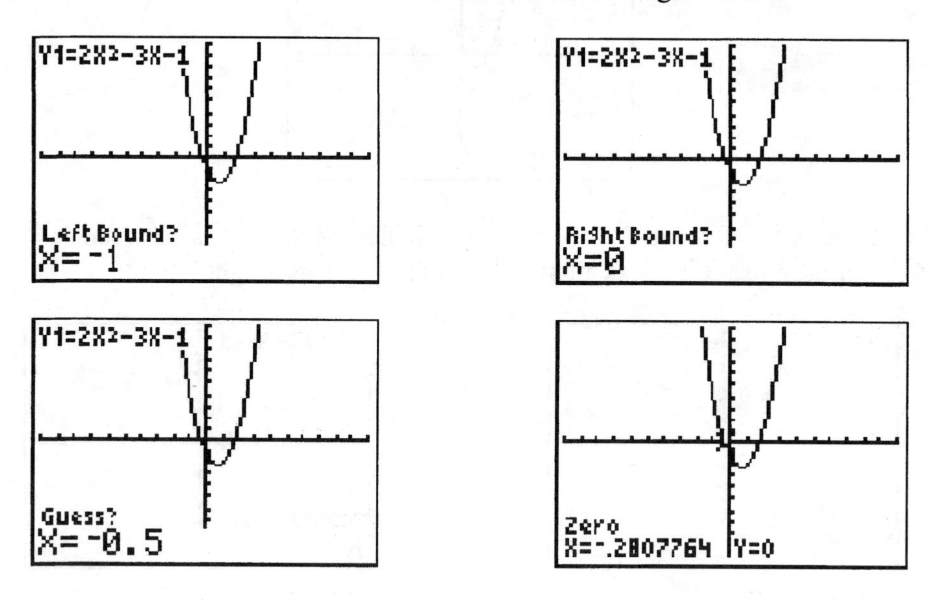

Thus, the smaller x-intercept is approximately $(-0.28, 0)$.

In Problems 7–14, use ZERO (or ROOT) to approximate the positive x-intercepts of each equation. Express the answer rounded to two decimal places.

7. $y = x^3 + 3.2x^2 - 16.83x - 5.31$

First, enter the equation in the function editor. Then, find a viewing rectangle that will show all the important features of the graph.

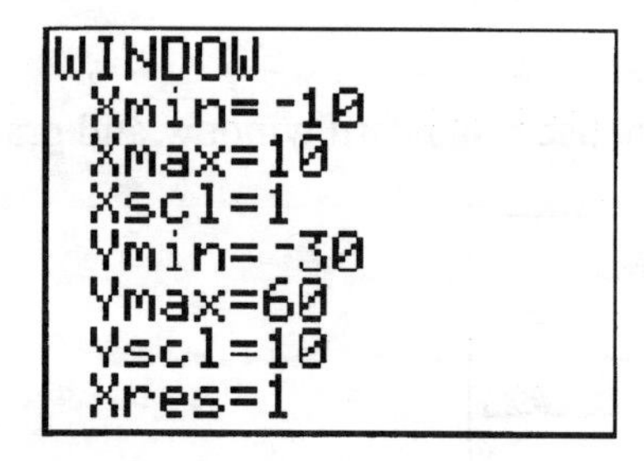

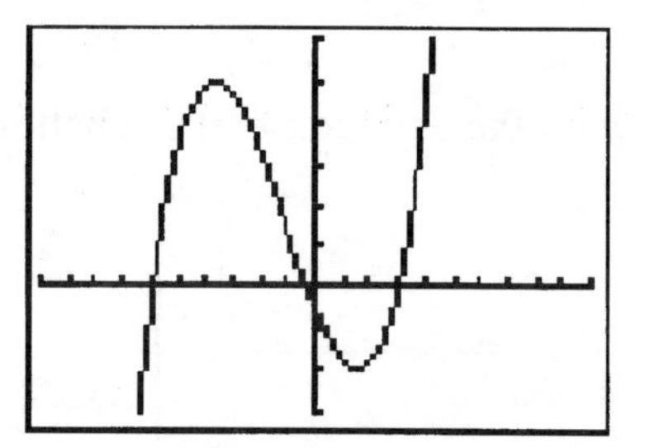

Find the positive x-intercept. Notice that the positive x-intercept is between $x=2$ and $x=4$, so we can use $x=2$ as a left bound and $x=4$ as a right bound.

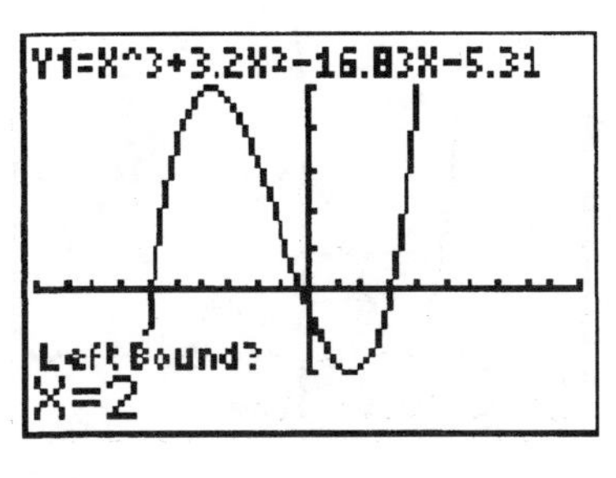

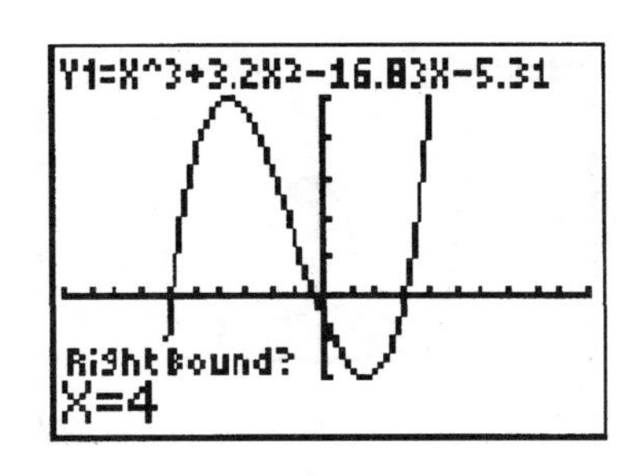

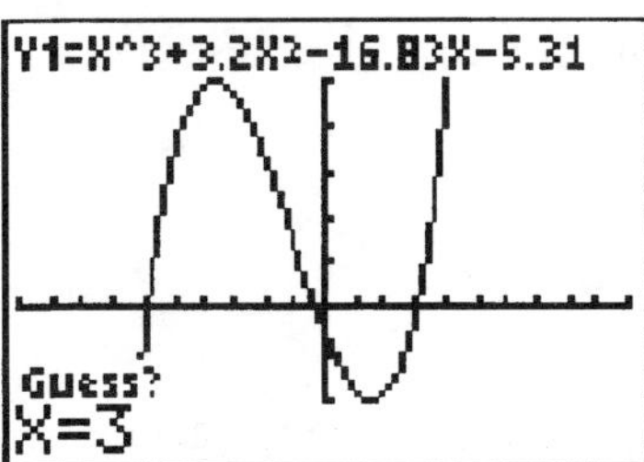

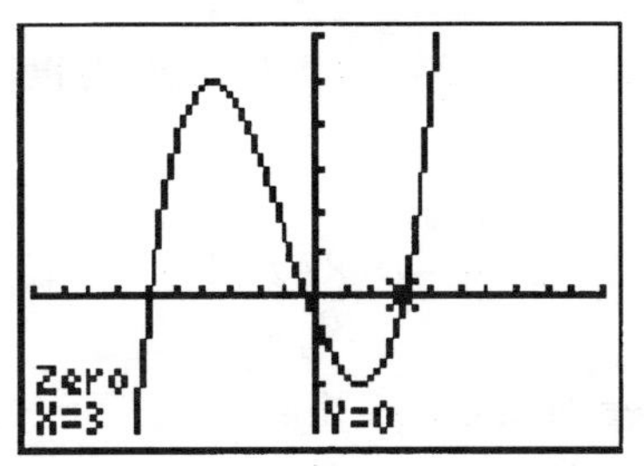

Thus, the positive x-intercept is $(3,0)$.

9. $y = x^4 - 1.4x^3 - 33.71x^2 + 23.94x + 292.41$

First, enter the equation in the function editor. Then, find a viewing rectangle that will show all the important features of the graph.

WINDOW
Xmin=-10
Xmax=10
Xscl=1
Ymin=-100
Ymax=500
Yscl=50
Xres=1

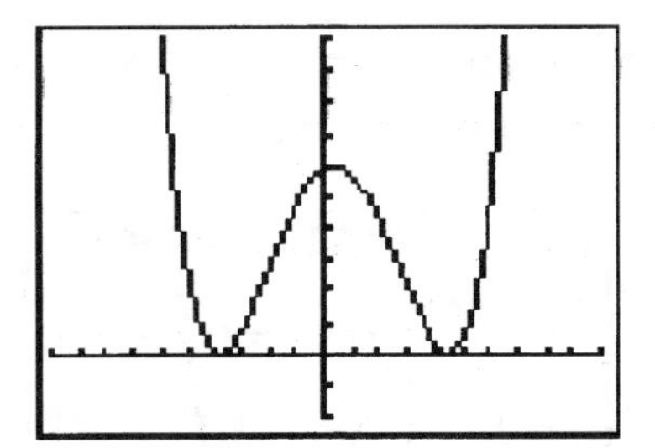

Find the positive x-intercept. Notice that the positive x-intercept is between $x = 3$ and $x = 5$, so we can use $x = 3$ as a left bound and $x = 5$ as a right bound.

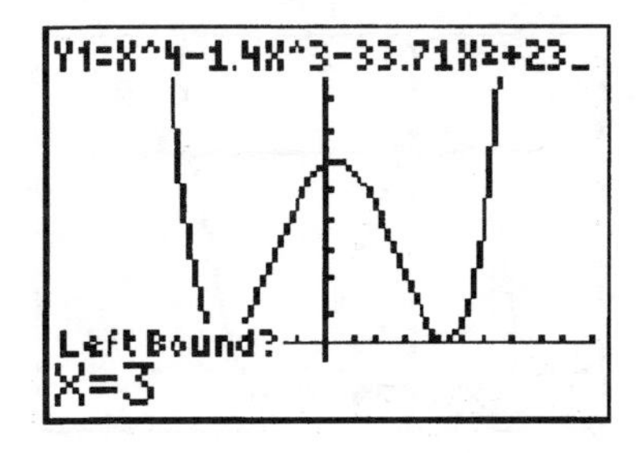

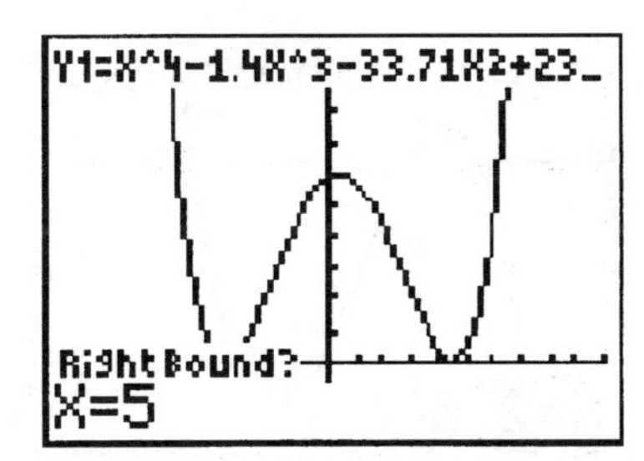

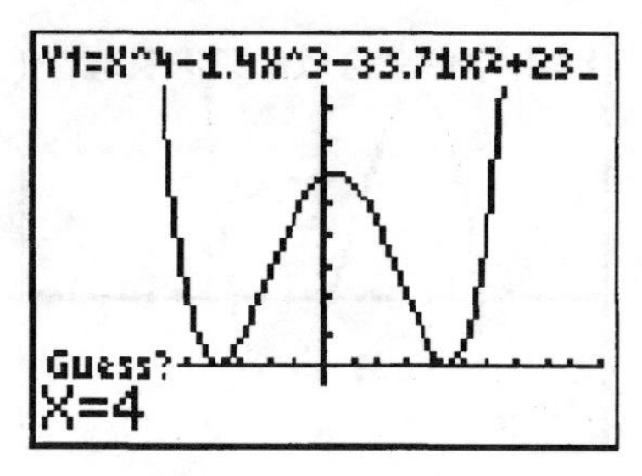

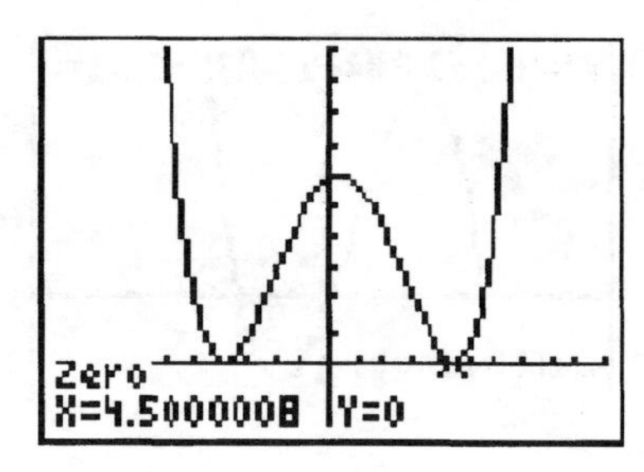

Thus, the positive x-intercept is approximately $(4.5, 0)$.

11. $y = \pi x^3 - (8.88\pi + 1)x^2 - (42.066\pi - 8.88)x + 42.066$

First, enter the equation in the function editor. Then, find a viewing rectangle that will show all the important features of the graph.

```
WINDOW
 Xmin=-10
 Xmax=20
 Xscl=5
 Ymin=-1250
 Ymax=500
 Yscl=250
 Xres=1
```

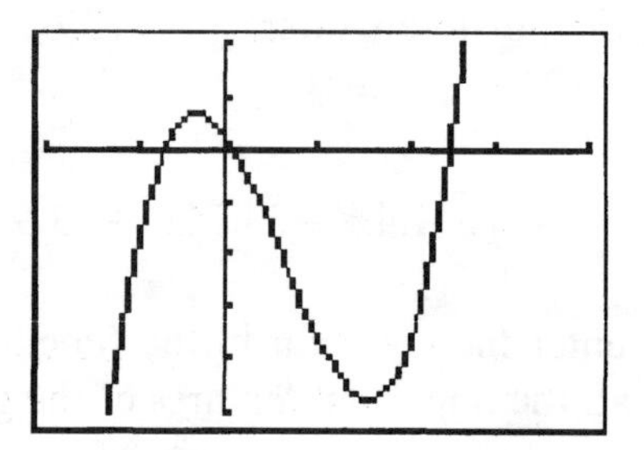

Find the positive x-intercepts. Notice that the first positive x-intercept is between $x = 0$ and $x = 5$, so we can use $x = 0$ as a left bound and $x = 5$ as a right bound.

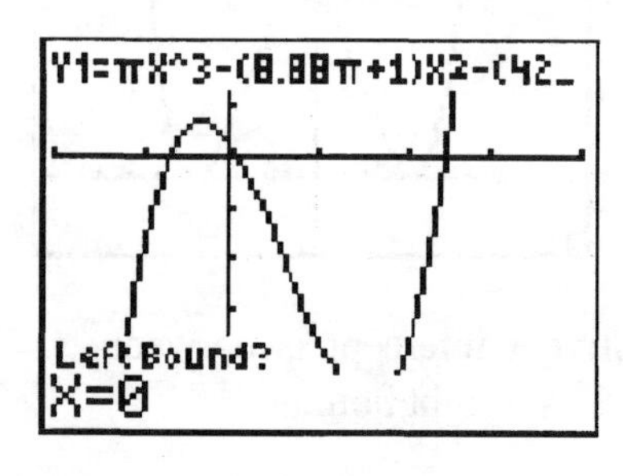

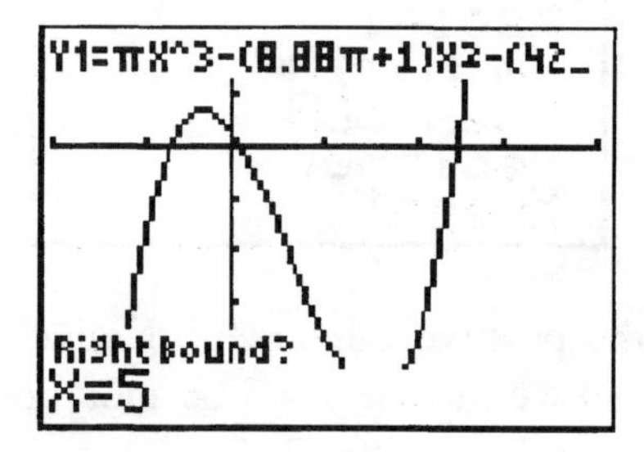

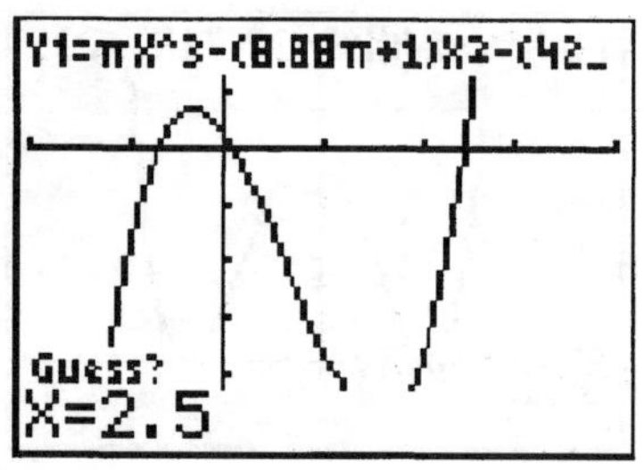

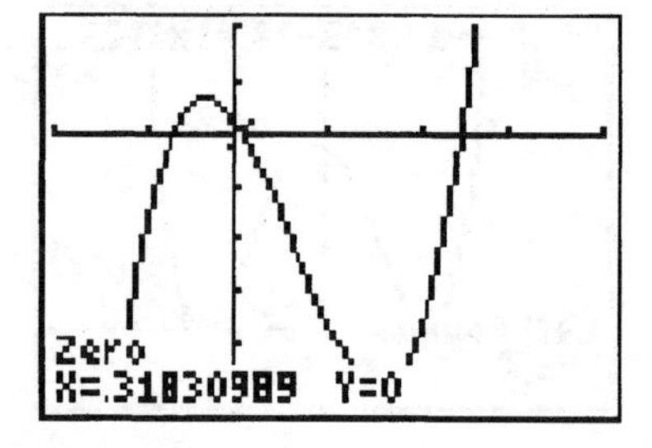

Thus, the first positive x-intercept is approximately $(0.32, 0)$.

Notice that the second positive x-intercept is between $x = 10$ and $x = 15$, so we can use $x = 10$ as a left bound and $x = 15$ as a right bound.

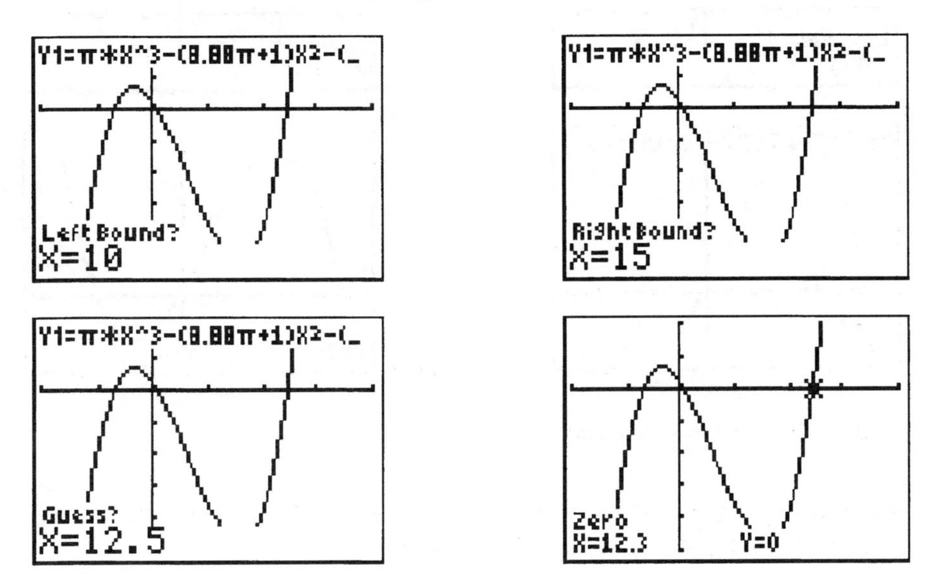

Thus, the second positive x-intercept is $(12.3, 0)$.

13. $y = x^3 + 19.5x^2 - 1021x + 1000.5$

First, enter the equation in the function editor. Then, find a viewing rectangle that will show all the important features of the graph.

```
WINDOW
 Xmin=-50
 Xmax=40
 Xscl=10
 Ymin=-10000
 Ymax=25000
 Yscl=2500
 Xres=1
```

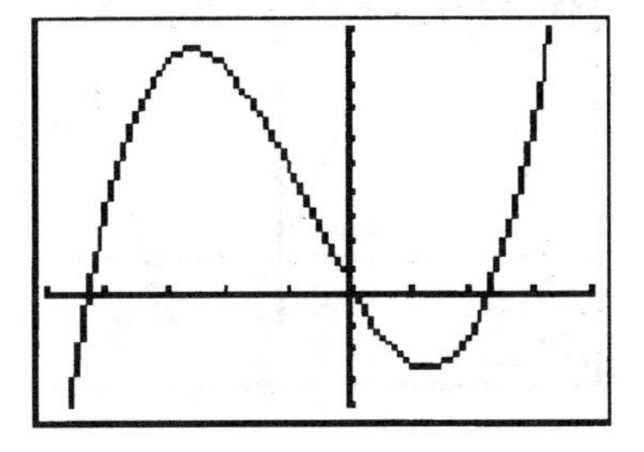

Find the positive x-intercepts. Notice that the smaller positive x-intercept is between $x = 0$ and $x = 10$, so we can use $x = 0$ as a left bound and $x = 10$ as a right bound.

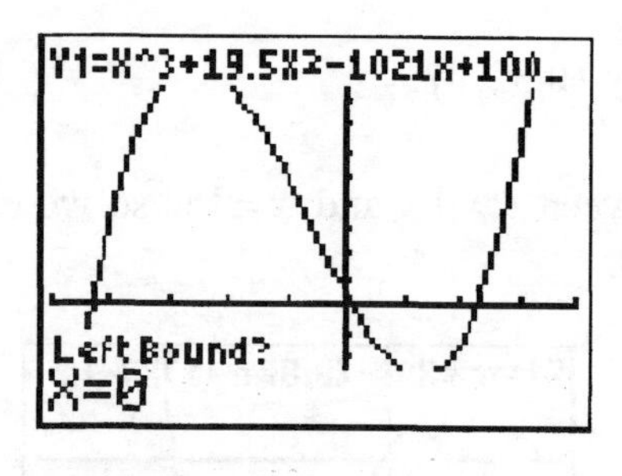

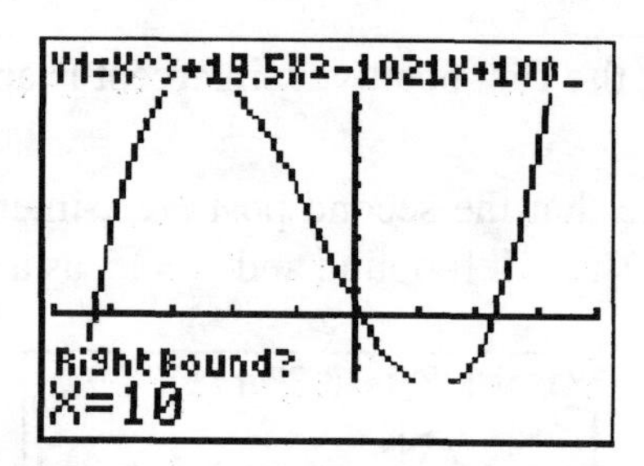

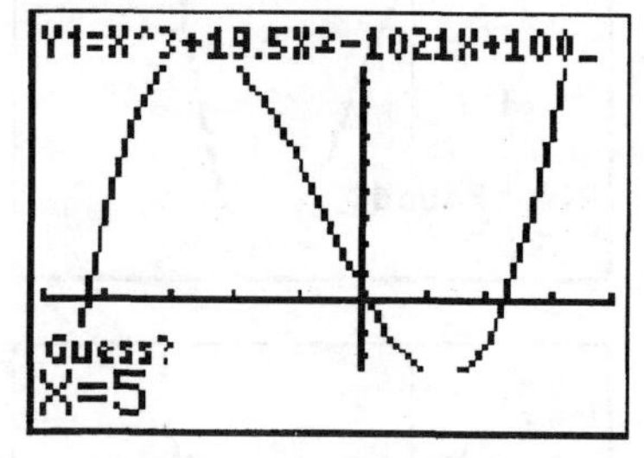

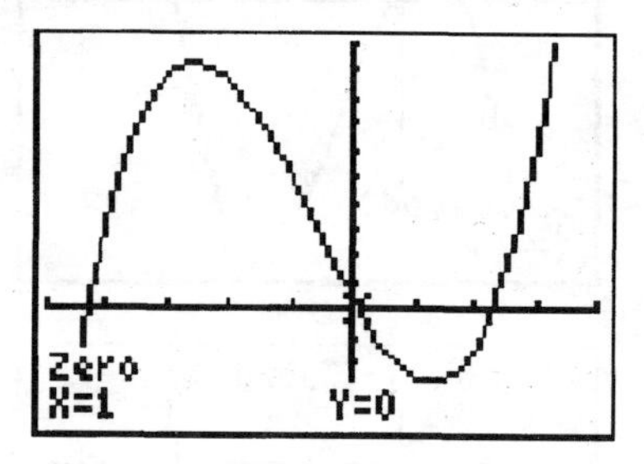

Thus, the smaller positive x-intercept is $(1,0)$.

Notice that the larger positive x-intercept is between $x=20$ and $x=30$, so we can use $x=20$ as a left bound and $x=30$ as a right bound.

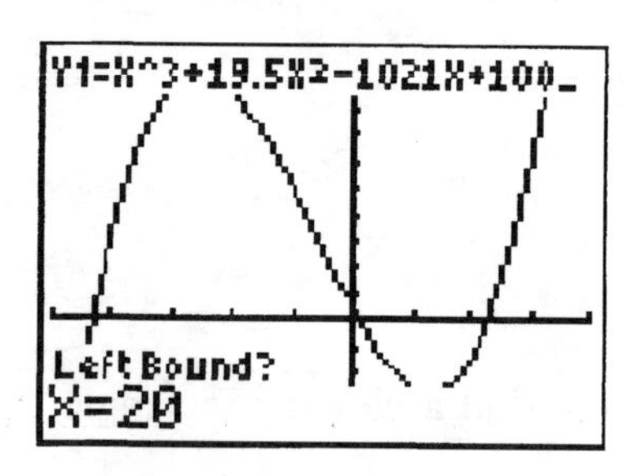

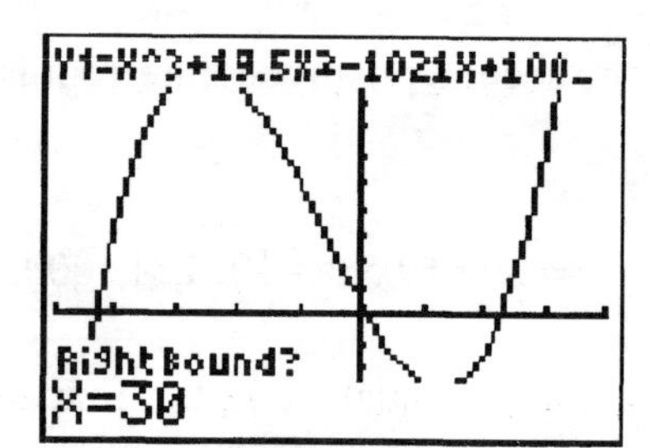

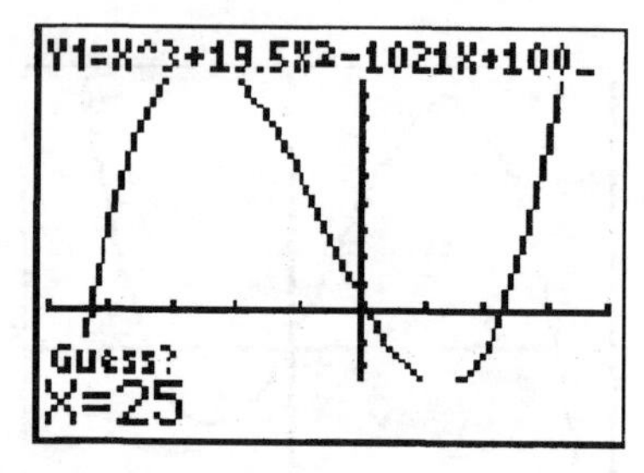

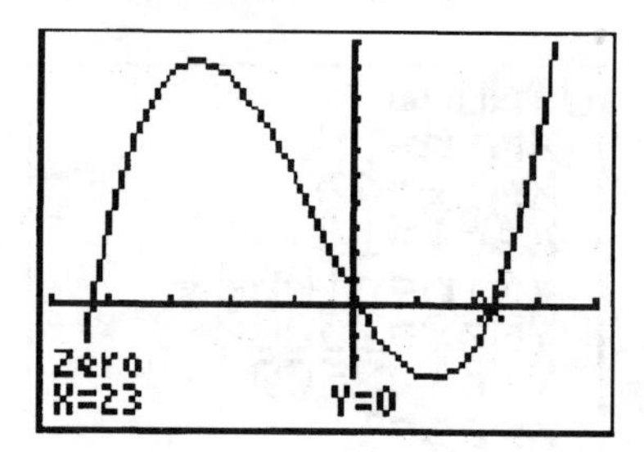

Thus, the larger positive x-intercept is $(23,0)$.

In Problems 15–18, the graph of an equation is given.

(a) *List the intercepts of the graph.*

(b) *Based on the graph, tell whether the graph is symmetric with respect to the x-axis, y-axis, and/or origin..*

15.

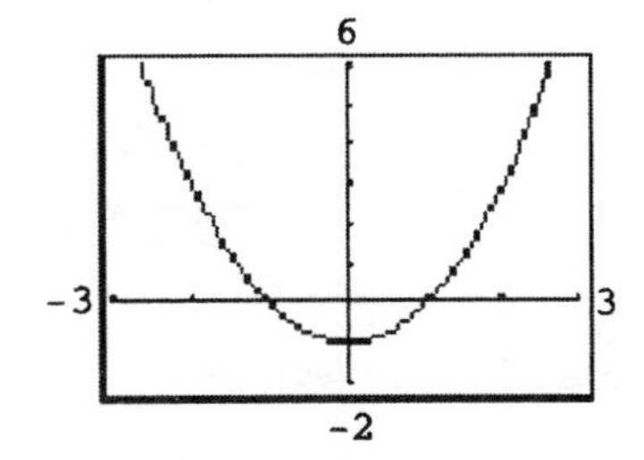

(a) The x-intercepts are $(-1,0)$ and $(1,0)$. The y-intercept is $(0,-1)$.

(b) The graph is symmetric with respect to the y-axis.

17.

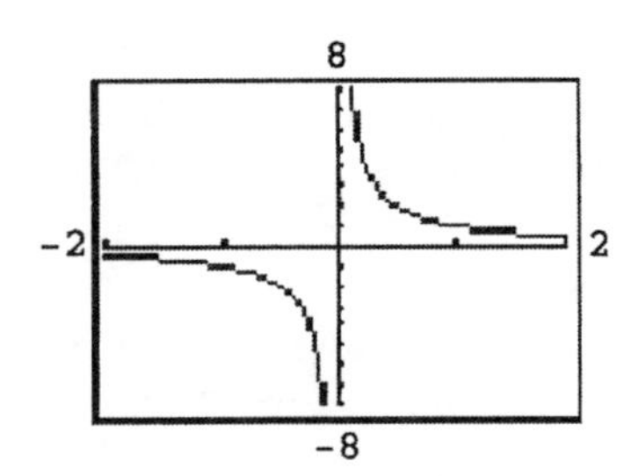

(a) There are no x-intercepts or y-intercepts.

(b) The graph is symmetric with respect to the origin.

Summary

There were no new commands introduced in this chapter.